Falling Glass

To Meg

Patrick Loughran

Falling Glass

Problems and Solutions in Contemporary Architecture

Birkhäuser – Publishers for Architecture
Basel · Berlin · Boston

We would like to thank the Chicago Committee on High Rise Buildings and the University of Illinois at Urbana Champaign for their generous support of this publication. Our thanks also go to the Francis J. Plym Traveling Fellowship.

Table of Contents

Foreword

page 7 Barometer, London

In the spring of 1999, I was the recipient of the Francis J. Plym Traveling Fellowship. This honor gave me the opportunity to study new building systems being developed in Europe that were not commonly found in the United States. As an architect with an engineering background, new technologies are of great interest to me. The topic of studies for the fellowship centered on innovation in building enclosure systems. I knew there was an architecture movement emerging in Europe that was based on unique technologies, and I sought to see at first hand how these great works were constructed.

The catalyst for this book and origin of the title came from a visit to London during my travels. I had read about a glass tower in the West End of London. The photographs in a book called "Sustainable Architecture and High Technology – Eco-Tech", by Catherine Slessor showed what looked like a building with a surreal blue glass enclosure. Looking at the pictures, I could not figure out how this skyscraper form was constructed. Completed in 1995, the Barometer was commissioned by the Thames Water Organization. It was intended to be a landmark on the eastern end of the Shepherd's Bush Borough. The architects designed the tower over an existing surge pipe. The tower acts like a giant operational barometer. A barometer monitor translates atmospheric pressure into an electrical signal that controls water jets inside the glass tower. As the pressure changes, so does the number of nozzles activated, causing the water level in the cylinder to rise and fall. Water discharged from the nozzles is collected and recycled via pumps and recirculation vessels at the base of the tower. The water is dyed bright blue so that the level can be seen more easily. At night the core, which is void, and the water were to be illuminated, highlighting the movement of the water.

When I arrived at the site, I was amazed at how small the Barometer was in reality. Positioned in the middle of a roundabout, the 15 meter tall element was not a skyscraper but more of a large glass sculpture. The photographs I had seen never contained anything but the tower, pointing toward the sky which made it difficult to gain a sense of scale. Looking up at the curved glass tube, I could see that the glass cylinder was composed of monolithic glass segments supported by through-bolt (counter-sunk) stainless steel fasteners. The glass edges were sealed from the exterior at butt joints. I was familiar with this type of point-supported glass detailing; however, I was interested in seeing how the details were executed on a curve. The inside face of the glass had a film of streaking blue color left from the dye in the water. Unfortunately, the Barometer was dry. I was disappointed to see that the water feature was not working, and the glass drum was empty. As I circled the tower to obtain a closer view, I saw a small sign, "DANGER - Possibility of Falling Glass", neatly taped on the stainless steel base of the tower. The message reminded me of the "Danger of Falling Ice" signs commonly found in the United States at the foot of tall buildings during winter months. Although all of the curved glass units were intact, there were fragments of broken tempered (toughened) glass on the ground around the tower.

Later that day, I spoke briefly with the architects regarding the history of the project. They confessed that the Barometer was not operational and had not been for months. When I asked about the broken glass, they explained

Blue dye is leaking out of glass connection. Barometer, London

that someone had thrown a brick at the sculpture. I uncovered no explanation for the sign at the base of the tower. I returned to the tower and reviewed the glass fittings with greater attention. Close inspection of the glass revealed blue dye oozing out of the countersunk fasteners. It appeared as if the fittings, common to the glazing industry (for flat glass panels), were allowing water to go from inside to outside. A designer should not reverse an enclosure system without investigating the ramifications to the details. We must remember that an umbrella may not work properly if held upside down.

This project initiated my interest in building enclosure failures. I started the fellowship with a quest to find out how the great contemporary masterpieces of Europe were constructed. I ended by trying to understand what went wrong with some of them. When I returned to

the United States, further research indicated that problems in building enclosure were not limited to European buildings or newly constructed enclosures. Many projects throughout the world over the past
70 years have struggled with glass enclosure issues; yet glass remains arguably the most common element in construction today. What contemporary building is constructed without the use of glass?

The projects discussed in this book were developed by some of the greatest architects of the 20th and 21st century. They are innovative, creative designs. With any new creative endeavor comes the possibility of flaws, and in some cases the "possibility of falling glass". The intention of the book is not to persuade for a return to more conventional building styles but rather to demonstrate how typical defects can be avoided. Glass is a unique construction material known for its transparency, yet plagued by its brittle nature. This publication describes many of the potential flaws of today's glass architecture and offers possible measures designers can take to prevent failures. In addition to pointing out problems, the chapters also examine enclosures that have been successful.

While this is a book about failures, it is not about liabilities, and the issue of whose fault these failures were remains deliberately undiscussed. The intention was not to determine responsibilites for past mistakes but to help avoid similar problems in the future.

Nickel Sulfide

Architects have used glass throughout history because of its transparent and stable nature. Modern techniques have improved the properties of glass; however, with improvements have come potential problems. Nickel sulfide is an impurity found occasionally in fabricated glass. It is during the melting process at the manufacturing stage that nickel combines with sulfur to form the nickel sulfide stone.
A nickel sulfide stone by itself is not harmful to the glass. It is after the glass is tempered (or toughened) through heat-treating that it can become harmful. Tempered glass containing a nickel sulfide stone is susceptible to a catastrophic glass failure. If the glass is exposed to a change in temperature, this change leads to a slight volume change in the stone. If the expanding stone is trapped in the center of the glass thickness it may cause a violent eruption commonly referred to as spontaneous breakage.
Depending on the size and location of the stone, the change can occur over a long period, usually years. Upon fracture the tempered glass will break into very small pieces. If the failure were frozen in place, the origin of the break could be traced to a small butterfly-shaped fracture. This pattern, also characterized as a "cat's eye" in the broken glass, centers on the nickel sulfide inclusion. The effect of nickel sulfide inclusions in glass have been known since the 1960s, but new techniques, and the increasing popularity of glass as a construction material, have brought the problem to wider attention.

An example of glass failure due to nickel sulfide inclusions exists in the International Terminal at Waterloo Station in London. The award-winning project was completed in 1994 to receive rail traffic from France through the Channel Tunnel link. This new structure reinvents the historic train stations of the Victorian

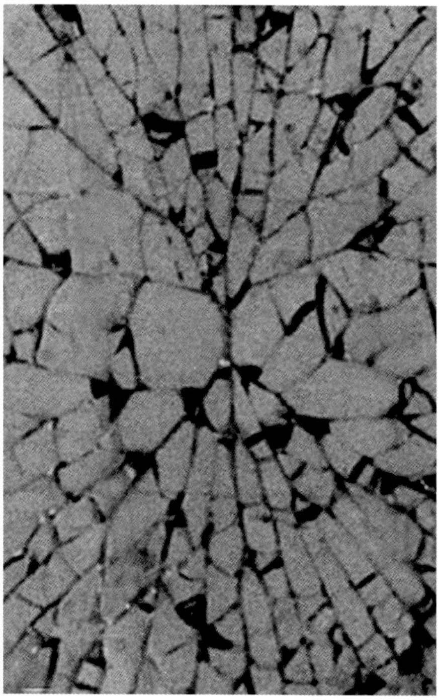

left The "cat's eye" pattern indicates that the tempered glass failure was the result of a nickel sulfide stone. The vertical line identifies the location of the expanded stone in the broken glass.

page 11 Butterfly-shaped fracture centered on nickel sulfide stone inclusion.

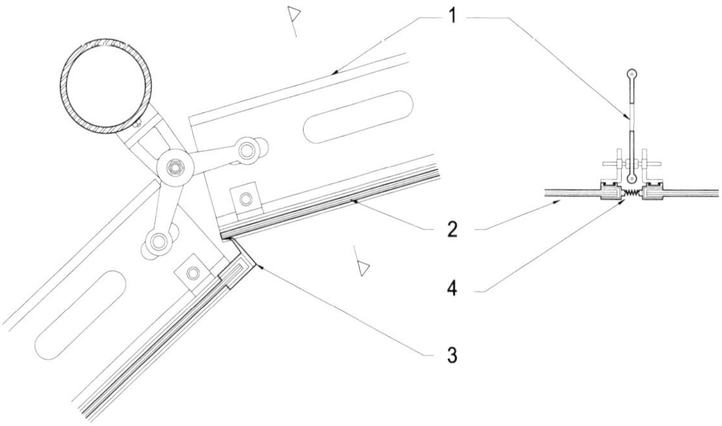

above An exterior view of the sophisticated skylight system. Waterloo Station International Terminal, London

below Sketch of skylight system: Standardized details permit independent movement of all monolithic tempered glass panes.
Waterloo Station International Terminal, London
1 Extruded aluminum skylight frame
2 10 mm tempered glass
3 Wiper blade seal at the tops of glass frames
4 Accordion gasket connection at the sides of glass frame

era. The grand hall gives a breathtaking experience of arrival to travelers visiting London by high-speed train. The long serpentine-shaped shed accommodates the 400 m long Eurostar trains passing in and out of the station and follows the predetermined line of the tracks. The train shed's basic structure is composed of a flattened three-pin arch composed of two bowstring trusses. Because of the asymmetrical geometry of the platforms, the center pin is moved to one side. West of the pin is a steeply pitched glass arch and east of the pin is a more gently curved arch composed of corrugated stainless steel decking.
An innovative skylight system was required to accommodate the complex geometry of the building. The glazing details are unique to skylight design. Monolithic 3/8 inch (10 mm) thick tempered glass panels are captured at the long sides in a metal frame. Accordion gaskets attach adjacent frames. A clip at the bottom supports the glass and a wiper blade seal at the top allows the glass to move independently from the glass above it. This segmented glass roof is very flexible and has been designed to accommodate the forces of wind, rain, and snow.

Shortly after completion, the project experienced glass breakage in the skylight portion of the vault. The glass skylight panels had been tempered but not laminated. Upon failure, tempered glass can barely sustain its own weight. In an overhead glazing application, it will fall out of its frame if not laminated. Broken fragments of tempered glass can be harmful to people below if they fall in a compact mass from great heights. A tent-like fabric beneath the glass portion of the Waterloo Station was added to catch the falling skylight glass like a basket. Overhead glazing regulations vary from country to country, and at the time of this project, it was not a requirement for the overhead glass to be laminated. The failure of the glass has been attributed to nickel sulfide inclusions. Glass breakage due to material contamination is an inherent risk of the tempering (toughening) process. In order to understand how the glass was contaminated, it is important to understand how the glass is made.

A screen below the skylight is used to protect pedestrians on the platforms from falling glass. The glass failure has been attributed to nickel sulfide inclusions.
Waterloo Station International Terminal, London

Raw Materials for Glass (The Batch)

The major component of glass is sand. This raw material is part of a complex chemistry used to produce a high-quality glass. The raw materials introduced at the beginning of the fabrication process consist of seven key components.

Nominal Composition of Clear Float Glass

Component	weight %
SiO_2 sand (quartz)	73.0
Na_2O	13.8
CaO	6.6
MgO	3.6
Al_2O_3	0.17
Fe_2O_3	0.12
SO_3	0.30

Nickel sulfide is not a component that is intentionally added to the batch mix for float glass. It is, however, the unwanted by-product of two elements that are inherent to the system.

Nickel (Ni)

Not shown as a raw material, nickel finds its way into the batch. The sources of the nickel contamination are not always known. It is important to note that the major component of the batch is natural sand from the ground. The sand can include small unwanted impurities. Undissolved inclusions may be incorporated within the float glass batch. These are broadly categorized as knots, dirt and stones. Most of these are considered imperfections but do not influence the performance of the glass. Industry standards provide allowable limits for size and minimum spacing of inclusions. Extremely minute powders or grains of nickel-rich materials can be found in these raw materials. In addition to tramp raw materials, nickel-rich contaminants, such as stainless steel, can be introduced into the glass batch from stainless steel hardware in contact with the mixture.

Sulfur (S)

Sulfur in the float glass process can come from two sources. A principle source of sulfur in the batch mix is sodium sulfate. Sodium sulfate is added to the glass batch to remove the bubbles or seeds in the mixture. A second source of sulfur can originate from the fuels used to melt the materials in the batch.

Nickel Sulfide (NiS)

The chemical creation of the nickel sulfide stones occurs in the molten stage of fabrication. Sulfur combines with tramp nickel to create the problematic inclusion. Nickel sulfide stones range in size from 3 and 15 thousandths of an inch (0.076 and 0.38 mm). Because of their small size, the stones cannot be seen through a practical inspection during

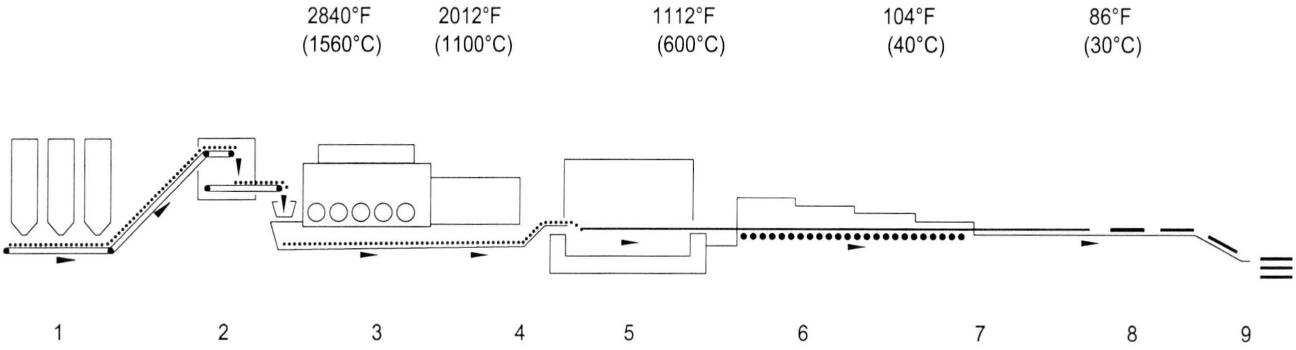

the float glass process. Currently, the float glass technology cannot totally eliminate the possibility of nickel sulfide stone inclusions in glass. Because the glass lights are not manufactured individually, but in batches, nickel sulfide is likely to be found in several lights; hence glass failures on buildings rarely occur with an isolated glass breakage. This type of failure often occurs on multiple glass lights on a single project. The failures can be traced back to a nickel-rich batch of molten glass.

The current method of fabricating flat glass is the annealed float glass process. Developed in 1955, this process consists of taking molten glass that has been floated on a bed of molten tin and cooling it slowly so that it contains no residual stresses. Annealed float glass can be cut without cracking beyond the cut line. When annealed glass breaks, it breaks in shards. For this reason, it is prohibited by most codes from use in doors and overhead glazing unless laminated.
It is important to note that nickel sulfide inclusions can exist in annealed glass and are not harmful to the material. It is only when these contaminates are combined with the secondary processing of tempering that problems can occur. Nickel sulfide stones are a very low-level impurity that would otherwise go unnoticed except for its occasional role in causing spontaneous glass fracture in tempered glass. A fundamental understanding of nickel sulfide failures comes from studying the tempering of annealed glass.

above Diagram of the float glass process. The nickel sulfide stone is an unwanted by-product of this system.
1 Introduce raw materials
2 Mixing
3 Melting
4 Refining
5 Molten tin float bath
6 Cooling
7 Cutting
8 Breaking
9 Storage

below Microscopic picture of a nickel sulfide stone (approx. 400 µm in diameter) from a window that experienced spontaneous breakage.

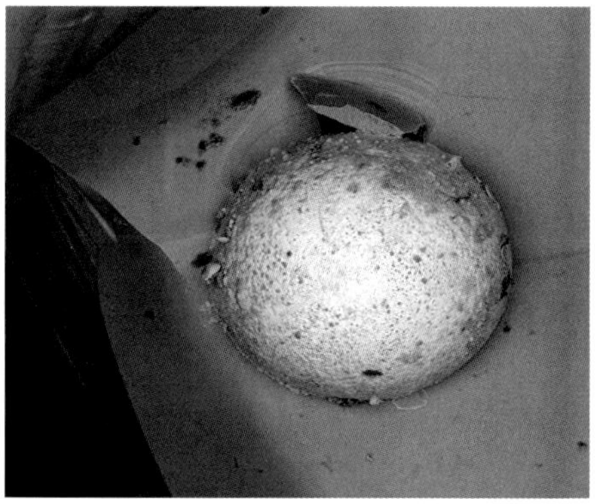

Tempered Glass Fabrication

In 1920, toughened glass was developed by the Société Anonyme des Manufactures des Glaces et Produits Chemiques de St. Gobain in Paris. In this process, annealed glass is sent through a furnace on rollers, heated to a semi-molten state, and then quickly cooled. The exterior surface cools more quickly than the inside of the glass, leaving the outer layer in compression and the core in tension. The tempering process creates a kind of thermal pre-stressing of the glass sheet. Because fully tempered glass has higher compression levels on the exterior surfaces, it usually yields a product that is four times stronger than annealed glass of the same thickness, size, and type. A single sheet of tempered glass cannot be cut, drilled, ground or polished after treatment. When broken at any point, by any force, the entire piece immediately breaks into very small granular pieces.

Spontaneous Breakage

When glass is tempered, the nickel sulfide inclusions are modified into a form that grows or transforms with time and temperature. Once glass is installed and the nickel sulfide inclusions are subjected to a change in temperature, small cracks may develop from the inclusion. If these cracks penetrate the tension layer of tempered glass, the resulting release of energy will cause the glass to break spontaneously.

Fully tempered glass is inherently and unavoidably subject to spontaneous breakage. This type of glass failure has been traced to the presence of nickel sulfide stones in the center tension zone of tempered glass. The fracture occurs when the stone undergoes a transformation accelerated by a temperature change. The temperature change leads to a small volume change in the stone (approximately 2 %). The volume change can lead to catastrophic failure. Because the glass breaks into small pieces, retention of the glazing in the opening is seriously compromised. Like magic, the enclosure can instantly change from a transparent barrier as strong as steel to a pile of crystals adjacent to an open hole. Why would such a material be used in construction? Ironically, the mode of failure of tempered glass is the source for its most common use: safety glazing.

Safety Glazing

Safety glazing is defined by the American National Standards Institute (ANSI) as follows: glazing material constructed, treated, or combined with other materials that, if broken, the likelihood of injuries that might result from such contact is minimized. In simple terms, safety glazing will not cut a person if it is broken. Safety glazing is used for transparent surfaces where strong impacts may be encountered and the hazard of flying glass must be avoided. Safety glazing requirements are designed to reduce or eliminate unreasonable risks of injury to humans when the glazed material is broken. Safety glazing is typically identified with a permanent label affixed to the glass stating it complies with safety glazing requirements. The following are five products that may be used in safety glazing applications in the United States while in most European countries only tempered and laminated glass are categorized as safety glazing. In Europe, the other three types may be used in selected safety applications.

Tempered Glass: a single sheet of specially heat-treated glass that breaks into small interlocking fragments when broken. It cannot be cut, drilled, ground or polished after treatment. When broken at any point, the entire piece immediately breaks in very small granular pieces. Fully tempered glass used in safety glazing applications is required to break into particles of a specified maximum size. If the particles are too big, the glass will not meet the safety requirements of safety glazing. Because the glass breaks in small pieces, retention of the glazing in the opening is seriously compromised. Fully tempered glass may fall from the opening in interlocking clumps.

A broken skylight panel with tempered laminated glass will not displace from the opening. Tape was added on the outside of this broken light for additional safety.
Illinois Center, Chicago

Laminated Glass: Two or more sheets of glass bonded together by an interlayer. Two common interlayer materials are plasticized polyvinylbutyral (PVB) and cured resin. Laminated glass has the added benefit of maintaining its form after breakage. For this reason, it is used in overhead glazing applications or where security is a priority. In overhead applications, it is not recommended to temper both lights of glass in a laminated panel. If both lights are laminated and both lights break, the glass panel will sag like a wet blanket and possibly displace from the frame.

Wired Glass: A single sheet of glass with wire completely imbedded in the glass but not necessarily in the center of the sheet. Wired glass is made by laying a wire mesh between two sheets of glass and hot-rolling the sandwich material. Wire is the traditional way of enhancing the fire resistance of glass. It can also be used as safety glazing because if the glass breaks, the wire will hold the sheet together.

Plastic Safety Glazing Material: a single sheet of synthetic plastic material, a combination of two or more such sheets laminated together, or a combination of plastic material and reinforced material in the form of fibers or flakes. Plastic sheets are more susceptible to scratching and discoloration than glass.

Organic Coated Glass: a sheet of glass coated on one or both sides with an applied polymeric coating, sheeting or film. This glass will break under sufficient impact, but the pieces of glass will adhere to the plastic and not displace from the sheet. This material can be added to existing glazing systems after installation. The addition of the film can be less expensive and disruptive than replacing the glass, however its durability after glass breakage is limited.

Local, state and national codes have adopted safety-glazing requirements. Codes vary; however, most define hazardous locations requiring safety glazing to include: doors, sidelights adjacent to doors, glass partitions, and overhead glazing. Some codes allow the use of a guard or "chair rail" to be used at full height glass enclosures in lieu of safety glazing. In locations where tempered glass is used as safety glazing, the risk of spontaneous breakage is present.

above Wired (not tempered) glass used on a window prevents falling glass after failure. Ridgeland Commons Ice Hockey Arena, Chicago

right Laminated (not tempered) glass used at exit doors shows how laminated glass is held within the frame after breaking.
Bibliothèque Nationale, Paris

right Plastic glazing is used on this entrance canopy in order to eliminate the risk of falling glass in this overhead glazing application.
Aéroport Bordeaux Mérignac

center The interior partitions of the upper floors of the bank were riddled with nickel sulfide failures shortly after installation.
Harris Bank, Chicago

below Sketch of tempered glass partition, Harris Bank, Chicago:
1 Steel tee support above ceiling
2 Lay-in acoustical ceiling tile
3 Aluminum frame fastened to tee above
4 1/2 inch (13 mm) clear tempered glass
5 Shim spacer and sealant separate glass from metal
6 Aluminum channel base fastened to concrete floor

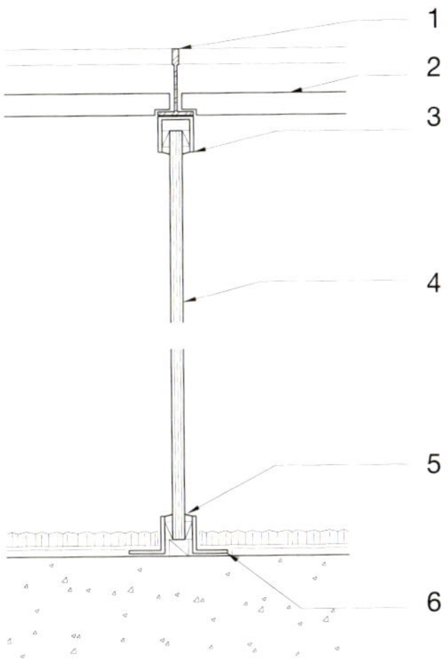

An example of a nickel sulfide failure in glass used for safety glazing is at the Harris Bank project in Chicago. In 1976, Harris Bank of Chicago was faced with glass breakage in their recently installed interior offices. The office partitions were composed of large tempered glass units 104 x 46 x 1/2 inch (2,642 mm x 1,168 mm x 13 mm). The partitions were used to form the walls of the interior offices in the building. The vertical edges of the glass were seamed, rounded and satin-finished. The glass lights were supported only at the top and the bottom edges with open joints at the sides. The glass used was tempered for safety reasons. The offices faced east with an unobstructed view of Lake Michigan and the rising morning sun.

The breakage of one light was observed by the office occupants. It was reported that after a period of exposure of the glass to solar radiation passing through the windows of the building, the office occupants closed the blinds on the interior side of the window and returned to the office. Shortly thereafter the tempered glass broke. This break exemplifies the classic failure of glass due to nickel sulfide. A total of six (10.7 %) of the 56 lights originally installed broke. What is interesting about the case is that it occurred in an interior glass partition. Apparently, with repeated exposure to solar radiation, significant stress had developed around the nickel sulfide inclusion to cause the glass to break.

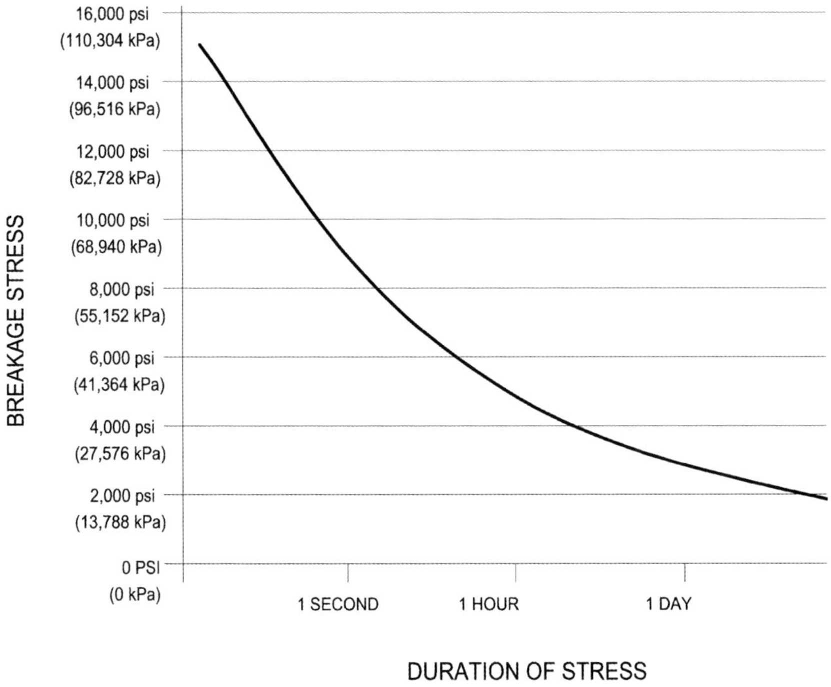

Annealed glass breakage stress as it is related to the duration of stress. Annealed glass is traditionally designed using a 6,000 psi (41,364 kPa) modulus of rupture (flexure) based on 60-second load duration.

It is important to note that there is nothing fundamentally wrong with the details used at the Harris Bank. Tempered glass has been installed in the same application in countless offices throughout the world without failure or incident. The occurrence of this type of failure in safety glazing is very rare. Tempered glass is used in most safety glazing applications throughout the world, and failures due to nickel sulfide stones are very rare. Tempered glass is also used where annealed glass is not strong enough to resist the design loads applied to it.

Glass Strength Properties

The strength of annealed glass diminishes with sustained load static fatigue. Traditionally, glass panel thickness has been chosen based on pane size and wind pressure using design tables developed by the glass industry. The strength properties of glass are determined at the time of fabrication. Glass strength properties can be improved through the tempering/quenching process.

In the design and use of glass for architectural glazing (windows, spandrel panels, and skylight applications), additional strength is sometimes required for wind loading or thermal stress breakage resistance. There are three methods of strengthening glass defined by the Glass Association of North America: chemical strengthening, heat-strengthening, and fully tempering:

Chemically Strengthened Glass

Chemical strengthening of glass is brought about through a process known as ion exchange. Glass is submersed in a molten salt bath at temperatures below the annealing range of the glass. At this point, an ion exchange takes place, which strengthens the surface to a depth of a few thousandths of an inch. Although chemically strengthened glass can be cut after treatment, the cutting process causes total loss of the added strength of the glass for an inch or more on either side of the cut. At these locations, it reverts to annealed glass. When chemically strengthened glass is broken, it breaks in a pattern similar to annealed glass and therefore should not be used by itself as a safety glazing material. The product is typically used in the ophthalmic and aeronautical industries where glass less than 1/8 inch (3 mm) thick is required to have properties stronger than heat-treated glass. It has also become popular in the security glazing field as a protective cover sheet over polycarbonates.

The entrance portal to the hotel suggests the anticipated movement of the glass façade with a trapezoidal stainless steel doorframe. The glass façade deflects in and out of plane; however, the stainless steel frame remains fixed. Hotel Kempinski, Munich Airport

Heat-Strengthened Glass

Heat-strengthened glass is annealed glass that is quickly cooled, so that the outer surfaces cool much more quickly than the core of the glass. This results in residual stresses with the surfaces in compression and the core in tension. This process, effective prestressing, makes the glass much stronger, approximately two times stronger than annealed glass. Because it breaks in a pattern similar to annealed glass shards, heat-strengthened glass does not qualify as a safety glazing material.

Heat-strengthened glass is produced with surface and edge compression levels less than fully tempered glass. The lower compression levels yield a product that is generally twice as strong as annealed glass of the same thickness, size, and type. Heat-strengthened float glass must have surface compression levels in the range of 3,500 psi to 7,500 psi (24,130 kPa to 51,700 kPa). Heat-strengthened float glass is typically produced in 1/8 inch (3 mm), 3/16 inch (5 mm), and 1/4 inch (6 mm) thicknesses. It is difficult to produce 3/8 inch (9 mm) heat-strengthened glass in the required stress range. Thicker glass (more than 3/8 inch or 9 mm) cannot be heatstrengthened on current equipment because of the large residual heat content as it emerges from the furnace.

Fully Tempered Glass

Tempered glass is similar to heat-strengthened glass but is made using a different temperature process resulting in higher residual stresses and greater strength. Glass is tempered by heating it to about 1,292° F (700° C), and then carefully cooling it using air blowers to cool the surface rapidly while the center is still viscous. The surfaces harden and are then put into compression as the center cools and contracts. Similar to pre-stressing, the tempered glass produces a balanced stress condition across the thickness of the glass with no overall load applied to the glass. Fully tempered glass has higher compression levels which usually yields a product that is approximately four times stronger than annealed glass and twice as strong as heat-strengthened glass of the same thickness, size and type. As discussed earlier, fully tempered glass is considered "safety glass" because it breaks in small pebble-sized particles, which are more prone to fallout than annealed glass, or heat-strengthened glass. United State fabricators typically offer fully tempered glass in thickness of 1/8 inch (3 mm) to 3/4 inch (19 mm).

Heat-strengthened and fully tempered glass cannot be cut, drilled, or edged after being heat-treated. Sandblasting, etching, or V-grooving, if done after the heat treatment process, will reduce the thickness of the compression layer and thus reduce the strength of the light of glass.

Because nickel sulfide stone formation can occur in the production of float glass, nickel sulfide may be present in annealed, heat-strengthened, laminated, as well as tempered glass. Because of very low residual stress in annealed glass, and carefully limited surface compression and corresponding center tension stresses in heat-strengthened glass, these types of glass are almost never susceptible to spontaneous breakage due to nickel sulfide stones.

Although the project has not had a history of glass failure, the atrium façade of the Hotel Kempinski at Munich Airport is a good example of tempered glass used for its structural

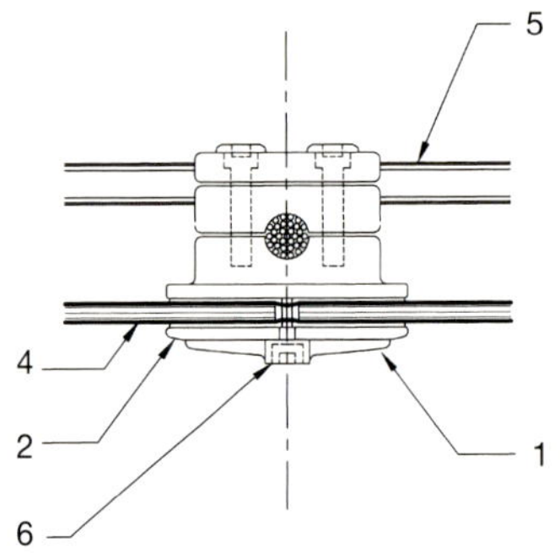

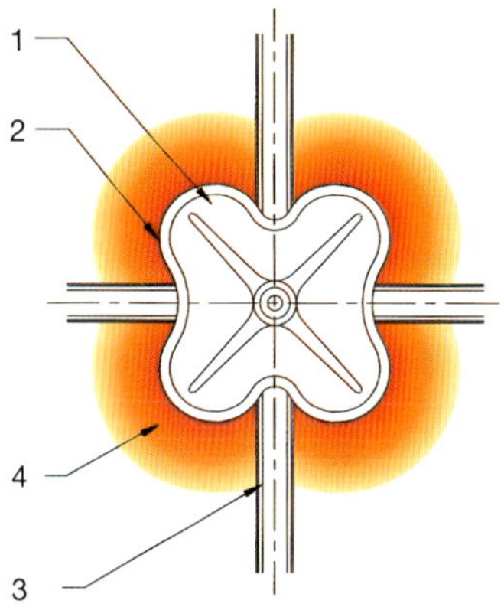

above left Section and plan of connection:
1 Cast stainless steel node
2 Silicone pads separate the metal from the glass
3 Silicone sealant separates individual panels
4 High stress levels on glass occurs at the connection
5 7/8 inch (22 mm) stainless steel cable
6 Joint between glass allows for a through fastener

above right The point-fixing glazing system requires tempered glass in order to sustain the design wind loads at the connections.
Hotel Kempinski, Munich Airport

page 21 The entrance to the 40 x 25 m glass façade required tempered glass to sustain the high stresses on the glass at the connections.
Hotel Kempinski, Munich Airport

properties. Completed in 1993, the atrium façades of Hotel Kempinski take the quest for transparency one step further. The project achieved a minimalist façade structure consisting of a simple cable net, 131 x 82 foot (40 x 25 m). High-grade stainless steel cables are used in diameter, spaced at 4 foot 11 inch (1.5 m). Large horizontal prestressed forces, 19 kips (85 kN), in the cables limit the deflection of the wall. Cable tensions will increase to 37 kips (165 kN) under full wind pressure when the maximum deflection is almost 3 foot (90 cm).

The point-fixing glazing system consists of cast stainless steel nodes at the corners of 3/8 inch (10 mm) thick laminated/tempered glass. The laminated layer improves the acoustical separation of the glass between the airport and the hotel lobby. The fastener does not penetrate the individual glass lights, because the fastener passes through the joint between them. The node clasp simultaneously connects the vertical and horizontal cables at the crossing points. The prestressing of the cables can be adjusted with threaded stainless steel fitting connections at the walls and the floor. A silicone pad inside and out separates the cast node from the glass. Because the forces on the glass at the connection are so great, tempered glass was required. The intersecting nodal shape appears to resemble the anticipated stress patterns on glass under loading. The atrium façades of the Kempinski Hotel are a courageous design effort that could not have been constructed without the use of tempered glass. The project has had no incident of nickel sulfide failure on these elevations, and probably never will. What measures are being taken on projects like this to reduce the incidence of nickel sulfide failure?

How can Nickel Sulfide be Avoided?

Greater Care at the Glass Fabrication Stage
In order to decrease the incidence of nickel sulfide stones, the glass industry has made considerable efforts in maintaining a higher level of care in fabrication. This includes minimizing contact of glass constituents and molten glass with nickel-bearing hardware, such as stainless steel. In addition, the raw materials and fuels used are closely inspected for contaminates.

Heat-Soaking
Heat-soaking of fully tempered glass is a procedure used to detect nickel sulfide contamination in glass prior to installation. In this procedure, sheets of tempered glass are exposed to extreme temperature variations prior to installation in order to identify and break contaminated lights of glass. Uncontaminated glass will not be affected by the procedure. During heat-soaking, the glass is subjected to elevated temperatures in order to accelerate the expansion in the nickel sulfide inclusions. Heat-soaking to test for spontaneous breakage varies from country to country. The process typically includes stacking the tempered glass in a walk-in, gas-fired air circulation oven. In many parts of Europe, the process is carried out in accordance with a DIN standard and the oven is raised to a temperature of 280 to 300° C (536 to 572° F) for an eight-hour duration. Oven sizes generally allow up to twelve stacks of glass in one test.

While heat-soaking does not guarantee that breakage will be eliminated in installed tempered glass, it is a valuable test for identifying glass batches which may have a high incidence of nickel sulfide inclusion. In a heat-soak program, glass is sampled from a lot or run on a statistical basis. During the statistical heat-soak procedure, 90 % or more of the glass that might have broken in the field will fail in the chamber. Because spontaneous breakage from nickel sulfide stones may occur even after fully tempered glass has been heat-soaked, some critics of the process believe that heat-soaking is not cost-effective.

Detection After Installation
Since 1991, the University of Queensland, Australia, has been working on a photographic method of detecting nickel sulfide stones in glass. The technique is used to detect nickel sulfide stones in existing glass façades exhibiting problems with nickel sulfide breakage. An apparatus is used to survey all external vision panels. This technique was successful in locating one hundred glass panels containing nickel sulfide on a 36-storey building in Brisbane, Australia. The panels were removed from the building after being identified in 1997. This method of nickel sulfide detection has proven to be 95 % effective.

Laminated Glass in Lieu of Fully Tempered
Where safety glass is required, laminated glass can be used as an adequate substitute for tempered glass. There can be a misconception regarding the safe nature of tempered glass. It is true that the glass breaks into small pieces that will not cut skin. However, if a large quantity of this material falls from great heights, it can be harmful to people below.

In overhead glazing, laminated glass is typically preferred and sometimes required by building codes. Laminated glass will crack and break under sufficient impact, but the pieces of glass tend to adhere to the interlayer and hold the glass in place. In unconventional overhead glazing applications where laminated glass is used (for example point-supported glazing systems), some municipalities require testing of a broken light of glass for fallout resistance. This precaution reduces the risk of a falling light of laminated glass.

Heat-strengthened Glass in Lieu of Fully Tempered

Heat-strengthened glass is far less susceptible to spontaneous breakage due to nickel sulfide inclusions. Current data suggests that surface compression levels in the range of 3,500 psi to 7,500 psi (24,130 kPa to 51,700 kPa) may eliminate any potential for nickel sulfide related breakage. Heat-strengthened glass typically limits surface compression levels to a range of 5,500 psi (37,920 kPa), plus or minus 1,500 psi (10,340 kPa). It should be noted that heat-strengthened glass is not a substitute for tempered glass in applications where "safety glass" is required.

This heat-soak chamber at a glass manufacturing facility in France presents a method of identifying nickel sulfide stones in glass prior to installation of glass on building.

Lessons Learned

1

In order to minimize the risk of failure due to nickel sulfide, alternatives to tempered glass should be explored. For example, heat-strengthened glass can be used to increase the strength of glass without tempering the material. Laminated glass can be used to meet safety glazing requirements without using tempered glass.

2

In overhead glazing systems, the bottom light of glass should be laminated to prevent displacement upon breakage. If point-supported glass is used in overhead conditions, some municipalities will require testing of the broken glass for fallout resistance. This precaution minimizes the risk of displacing a light of laminated glass in an overhead application.

3

Nickel sulfide may be a problem in tempered glass; however, nickel sulfide failures are not limited to exterior glazing applications. Interior glazing exposed to direct sunlight can acquire the thermal change required for spontaneous breakage.

4

If tempered glass is required, it may be prudent to heat-soak the glass prior to installation. Because nickel sulfide contamination is linked to nickel in the batch ingredients, glass breakage is likely to occur in several pieces of glass from the same batch. Heat-soaking the glass may not identify all the contaminated glass in a batch. However, it may be used to alert glass manufacturers to a possible contaminated batch in their process. Techniques are being developed to detect nickel sulfide in glass that has been installed.

5

Because of improvements in the quality control of the glass manufacturing industry, the probability of nickel sulfide contamination has been greatly reduced; however, it has not been completely eliminated.

Thermal Stress

If a baseball is found on the floor surrounded by glass fragments, no great analysis is required to determine the cause of the broken window. The source of glass breakage on a building is not always so obvious. Nonetheless, the cause of the failure can often be determined with equal certainty by studying different types of glass failures and how to detect them. One type of glass failure that can be identified through close examination of the broken glass is thermal stress. Many people have experienced a thermal stress fracture by mistakenly placing a very hot china dish in cold water. Under these extreme conditions, the dish can break. The origin of the break can often be traced to a chip in the dish perimeter.

A similar phenomenon frequently occurs in architectural glazing. The catalyst of thermal stress fracture is again heat. Sources of heat in building glazing can vary. Interior heating systems and fire are two causes, but the most common source of heat on glass is the sun. Like most materials, glass will expand when exposed to heat. If the heat is uniformly applied to a piece of glass, the result will be the expansion of the entire panel. However, if the glass is not uniformly exposed to the heat, the warm section will expand and the cold section will remain relatively static. The cold section will resist the expansion of the heated portion. This resistance causes tensile stress in the glass. The amount of stress depends on the degree of temperature differential and the level of heat absorption by the glass.

More likely than not, the highest stress and origin of fracture will occur at the glass edge. This is true for two reasons. First, the edges of glass windows are normally recessed in a glazing rabbet and adjacent to the large thermal mass of the building. Protected from the sun, shaded behind a glazing cap, the

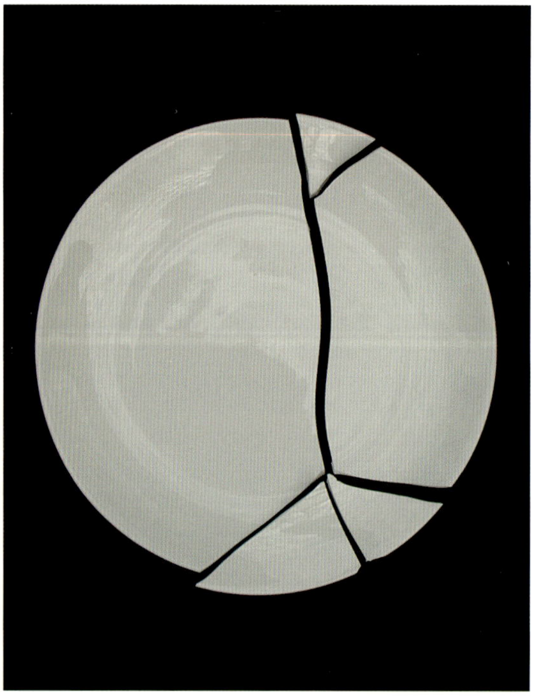

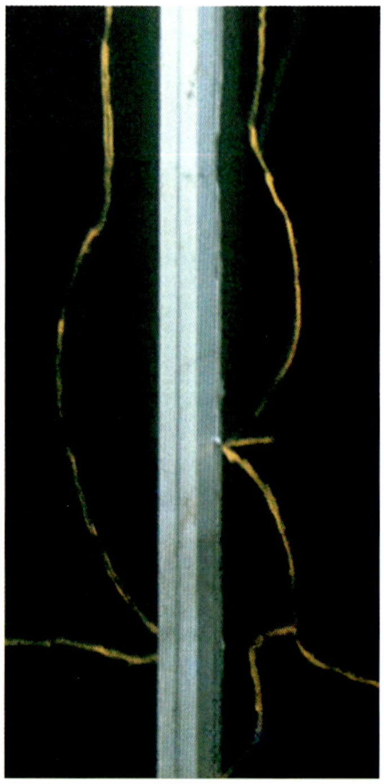

left Thermal stress fracture can occur if a very hot dish is placed in cold water.

right Thermal stress cracks developed in a tinted glass panel exposed to extreme heat.

page 25 The curtainwall's red color is a distinctive element in the Chicago Skyline. The façade has had a history of glass failure attributed to thermal stress.
CNA Building, Chicago

right The textured façade casts shadows on the glass windows.
CNA Building, Chicago

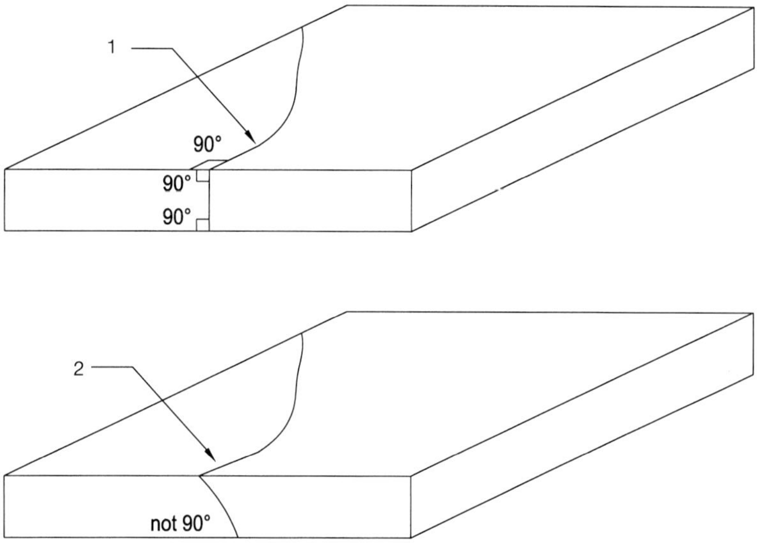

above Identifying the cause:
1 Thermal breaks in glass can be identified by the line of the crack being perpendicular to the edge surfaces.
2 Non-thermal breaks can be identified by lines that are not perpendicular to the edge surfaces.

glass edge remains much cooler than the rest of the glass pane. Secondly, the edges of glass frequently contain minor flaws, due to the cutting and shipping of the panes prior to installation. These flaws provide a weak point in the glass that is susceptible to the build-up of thermal stress. When the force of thermal stress pulls equally on both sides of the defect, the crack generated from the force will be at a right angle to the surface of the glass. This type of fracture can be identified by a break, which is perpendicular to both surfaces of the glass edge. There are many factors that contribute to thermal stress in glass. All glass experiences thermal stress and can be designed to withstand it. The probability of glass fracture due to thermal stress has been greatly reduced with the use of heat-strengthened or fully tempered glass. Nonetheless, an understanding of this type of glass failure is important in order to avoid thermal stress failure by identifying locations where heat-treated glass should be used.

An example of glass failure due to thermal stress exists in the CNA Building in Chicago. Completed in 1973, the building became an instant landmark, receiving a citation of merit in 1974 from the Chicago Chapter of the American Institute of Architects. The 45-storey skyscraper has a red color (nearly the same shade as San Francisco's Golden Gate Bridge) that made it a distinctive building in the Chicago skyline. The glass windows are set back from the red steel frame by a 1 foot 3 inch (40 cm) projection. The façade was praised for its strong play of light and shadow. Ironically, the façade's ability to play with the light was one of the contributing factors to a history of glass failure. Thermal stress has been determined to be the cause of the window cracking.

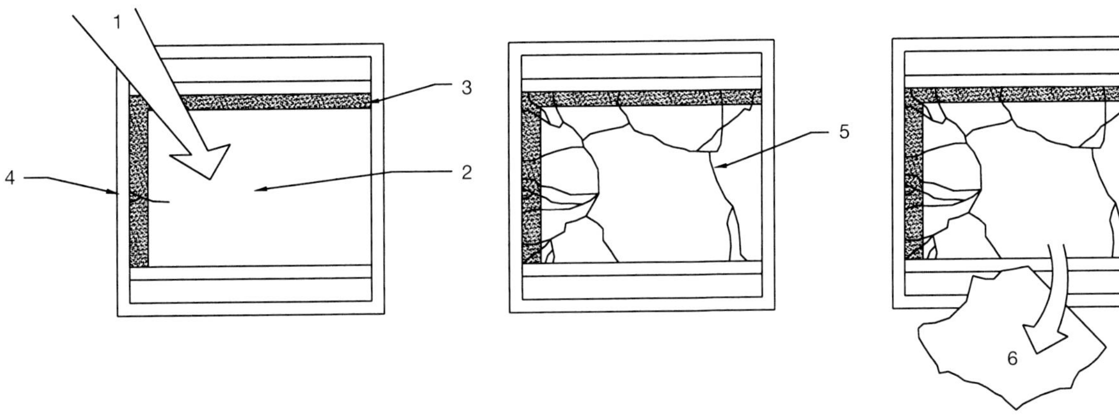

The progression of thermal stress failure at the CNA Building, Chicago:
1 Sunlight heats up the glass exposed to view.
2 The center portion of the glass expands due to heat.
3 The glass edges remain cool in shadow and resist the expanding forces of the center.
4 Thermal stress cracks originate at the edges which are the weakest part of the glass.
5 As stress continues, with contributions from wind, the cracks expand.
6 The glass fragments can come loose from the pane.

The thermal stress that caused the cracks was exacerbated by a combination of factors. The annealed glass panes were not clear but tinted gray. The darker colored glass absorbed much more heat than clear glass. The glass panes were oversized and thicker than normal window units: approximately 6 foot 7 inch high (2 m), 9 foot 6 inch wide (2.9 m), and typically 3/8 inch (10 mm) thick. 1/2 inch (13 mm) thick glass was used at the building corners to resist higher wind loads. The windows create large, thick surfaces of glass to collect the sun's heat. The 1 foot 3 inch (40 cm) recess at the perimeter of the window created non-uniform shading on the glass at certain times of the day. This type of shading is a classic source of thermal stress, contributing to the cooler temperature at the perimeter. Heat trapped by the venetian blinds on the interior of the windows contributed to the failures by heating the center portion of the glass material.

Many of the building's glass windows started cracking shortly after the building was opened. More than 100 panes of glass broke from 1972 to 1974. All of the cracks were found on the building's south, west and east sides. These façades receive the most direct sunlight. A proposed solution to the problem was to reinforce the glass with a heavy polyester laminate film bonded to the inside surface of the window with acrylic adhesive. The added film was intended to hold shards of glass in place if the window broke. The film repair only represented a temporary fix after the glass had broken. Once a window is broken, high winds can tear the film and dislodge portions of the window from the building. The CNA curtainwall depicts many of the common causes of thermal stress fracture in glass façades. At the origin of this type of glass failure is thermal expansion.

Thermal Expansion

All materials expand and contract when exposed to changes in temperature. For some materials, the change is great. Aluminum expands twice as much as steel and concrete. Acrylics expand more than three times as much as aluminum. There is a simple formula for predicting the change in length of an object exposed to a change in temperature. An object of initial length L at some temperature will change in length ΔL if the temperature changes ΔT.

$$\Delta L = \alpha \Delta T L$$

The constant α is called the average coefficient of linear expansion. This coefficient varies from one material to the other.

Coefficients of Expansion α

Material	inch/inch/degree F x 10^{-6}	cm/cm/degree C x 10^{-6}
Acrylic Plastic	40	72
Polycarbonate Plastic	38	68.4
Aluminum	13	23.4
Bronze (Arch 385)	11.6	21.9
Copper (110)	9.8	17.6
Stainless Steel (type 304)	9.6	17.3
Structural Steel	6.7	12.1
Concrete	6.0	10.8
Glass (float)	5.1	9.2

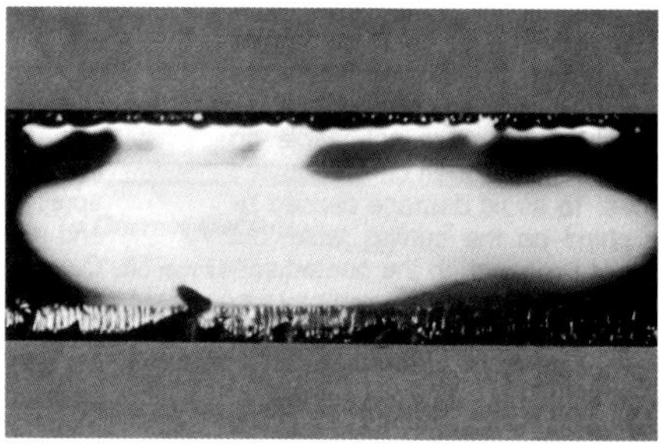

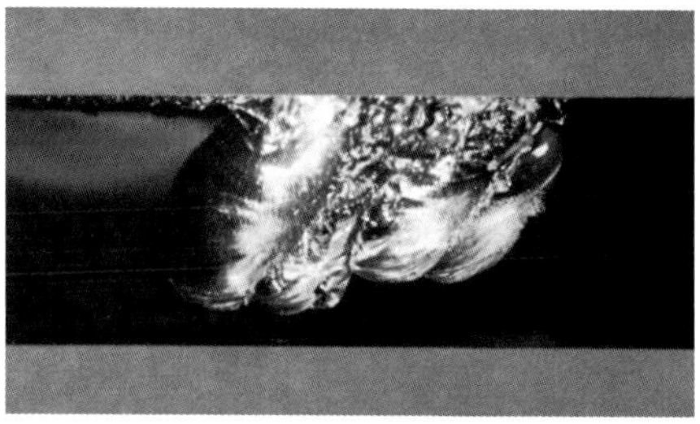

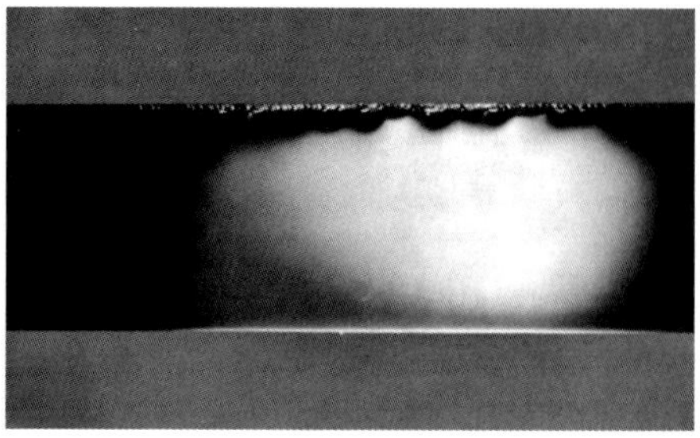

left, from top to bottom
Deep shark teeth (unacceptable)

Serration hackle with spalls (unacceptable)

Rough chip (unacceptable)

right Clean edge (ideal)

Imperfect Edges

Most people understand that an imperfection in the edge of a material is a point of weakness. When we try to open a bag of chips, our eyes search for the serrated edge, the imperfection in the seam. We know from experience that this is the easiest location to tear the bag open. The sawtooth edge represents a weak point in the enclosure and the best starting point to tear. This everyday experience helps us to understand where thermal stress cracks start.

When a material is exposed to stress, the forces cause the most damage at the weakest points. The weakest point for glass is at defects in the glass. Defects can be categorized into three groups: defects due to internal glass flaws like inclusions of foreign material from the batch and melting process, defects from the fabrication process like the contact between hot glass and colder metal, which precipitates small fissures in the glass, and lastly, mechanical damage to the glass after fabrication. This last group is very common and can often occur when the glass is cut. After glass is manufactured in large sheets from the float glass process, it must be cut to the specific sizes required for the project. Once the glass has been cut to size, the edges are inspected for defects and eased in order to aid handling. Any minor chips are smoothed out. If undetected, a defect in the glass can be the starting point of cracks due to thermal stress. Edges can be damaged during manufacturing, packaging, transporting, handling or installation. The following are common edge defects to watch out for.

Glass expands and contracts at a lower rate than concrete, steel or aluminum, yet it is responsible for one of the most common modes of thermal failures in building enclosure systems. It expands when it gets warm and contracts as it cools. Typically these differential movements are so small that they do not have any effect on the glass. Two types of problems can occur from the differential movement of glass. If the glass is restrained from movement, stress will develop and can break the glass. Glass panes must be allowed to expand. Its movement should not be restricted inside the frame. The second problem can occur when one area of a glass pane gets warm and another area of the same pane remains cool – for example, when the center of a glass light receives direct sunlight and the edges remain in shadow. The warmer area of the glass, which wants to expand, begins to push against the cooler area, which is not expanding. The result of the non-uniform thermal expansion is thermal stress. Because of glass's relatively low fracture toughness, it is increasingly susceptible to failure when the material contains edge defects.

right The steel columns and frame supports cast shadows on the glass during certain times of the day. Bacardi Office Building, Mexico City

below The weakest point: similar to edge defects in glass, the sawtooth edge of a bag of chips represents the weakest point and the easiest place to start a tear.

Shark teeth are dagger-like imperfections that start from the scored surface of cut glass. Breakage possibility increases as the depth, roughness, and number of shark teeth increase.

Serration hackle is caused by edge imperfections, usually perpendicular to glass surface, which occur at surface opposite score. Breakage possibility increases as density and depth increase.

Rough chips or penetrating chips are often caused by impact damage.

It is an error to cut glass in the field prior to installation. Field tools can leave small cracks and frayed edges that can lead to failures of the units. An example of thermal stress failure due to edge defects occurred at the Bacardi Office Building in Mexico City. Completed in 1961, the two-storey administration building forms the entrance to the Bacardi Company's rum distillery. This low-rise building has a skeleton-like steel frame that has been painted black. A 30 foot (10.4 m) column grid supports the floor, roof, and curtainwall. The project depicts a style perfected with a minimalist "less is more" philosophy. Projects like this were revolutionary for their time and challenged society to think of buildings differently. The building design also required contractors to rethink how to construct buildings. This new international style of architecture used large lights of tinted monolithic glass recessed in dark steel frames. The glazing details for this type of building were susceptible to thermal stress cracks at the perimeter.

As a result of poor fabrication or installation, the glass of the Bacardi Office Building was installed with imperfect edges. Apparently during the field glazing process, many of the monolithic glass units were crimped with an instrument similar to pliers in order to fit the large lights of glass in their steel frames. The edge defects were undetected after installation because the glass edge was concealed behind the steel trim. The projecting steel window frames and structural columns cast shadows on the perimeter of the large glass panels during certain times of the day. Mexico City experiences cold nights at certain times of the year. In the morning, the rising sun hit the east elevation and the thermal stresses prompted by the change in temperature caused the glass to break. A review of the broken glass revealed jagged edges as the origin of thermal stress failures.

right The jagged edges of installed glass may have been the result of poor installation procedures. Bacardi Office Building, Mexico City

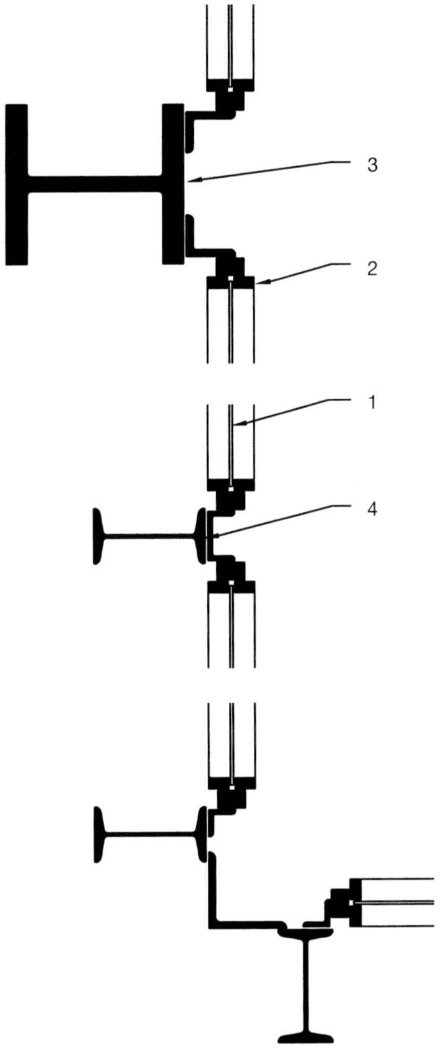

above Nipped or damaged edges can lead to thermal stress cracks.

right Sketch of plan detail, Bacardi Office Building, Mexico City:
1 Monolithic tinted annealed glass
2 Steel frame
3 Steel column
4 Steel support for frames

Fracture Patterns

Fractures in glass typically originate at the edge. It is possible to determine the type of fracture and the stress level associated with the crack by reviewing the fracture pattern. A single fracture line starting at an edge extending part way or all the way to another edge is described as a low stress break. Low stress breaks are normally considered to be less than 1,500 psi (10,335 kPa). Low stress breaks indicate that the origin of the fracture most likely came from a damaged edge. If a break begins in a single line, and within approximately 1 inch (24 mm) branches out into two or more fractures, the break is characterized as a high stress break. The branching indicates a high energy level, often due to high temperature difference between the center portion and the edge of glass. The closer to the edge that the branching occurs, the higher the stress and the greater number of branches. The maximum thermal stress occurs in the central half of the glass, decreasing to zero at the corners. Thermal breaks due to solar exposure would not be expected to occur within 12 inch (30 cm) of the corner. Failures located near the corners are most likely due to bending stress (pressure or impact).

Fracture patterns:
1 Low stress break (low energy level)
2 Low stress break (moderate energy level)
3 High stress break (high energy level)

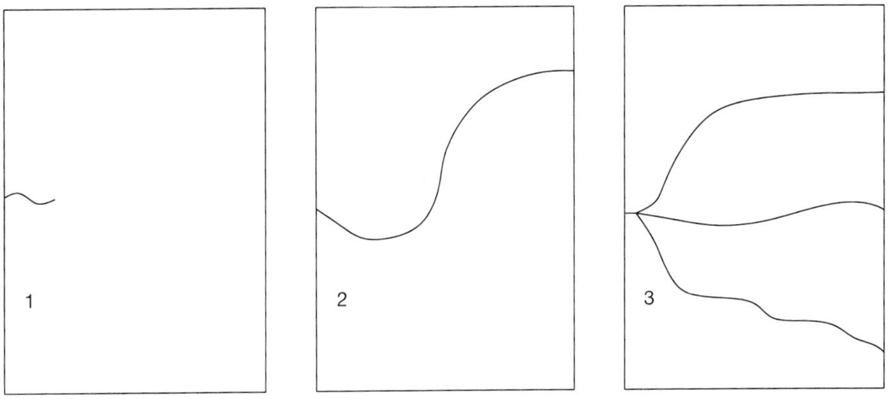

Contributing Factors

All glass experiences thermal stress; however, a combination of contributing factors can lead to greater levels of stress and glass breakage. In addition to imperfect edges, there are several factors that contribute to thermal stress fractures. Any factor that encourages an increase in the hot center/cold edge condition will contribute to thermal stress in glass:

Orientation of Windows: In the northern hemisphere, windows facing north are exposed to little direct sunlight. Façades facing east, west and particularly south can receive direct sunlight that contributes to thermal stress on glass.

Cold Weather Conditions: Although thermal stress failures are typically associated with solar radiation, a similar effect can occur during the dark hours of the night. Glass breakage can occur in cold climates, when there is a rapid decrease in outside temperatures, coinciding with high internal heating temperatures. In this case, the exposed area of the glass is kept relatively warm on the inside surface by direct heat, yet the outside air has cooled the window frame and consequently the glass edge. The result can be excessive stress and failure. The condition is particularly aggravated when the heated room side air is allowed to impinge on the glass surface. This is more likely to happen on the inner glass light of a double-glazed unit, when the internal heating keeps the glass warm but the glass edges are cooled by the conductive heat loss through the frame. Providing sufficient thermal isolation between interior and exterior framing elements can reduce the risk of this type of failure.

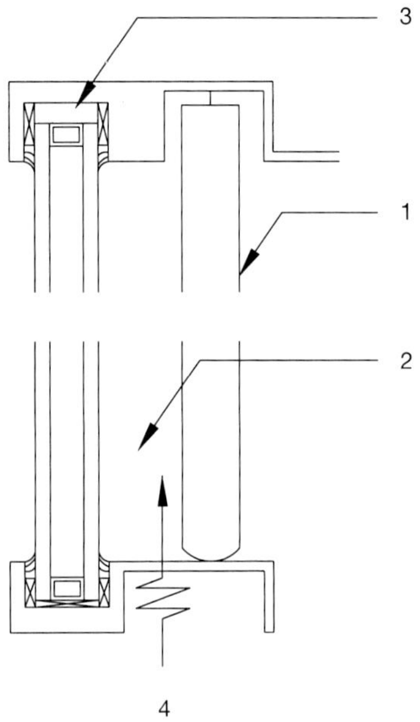

Shading device:
1 Internal shading devices can contribute to thermal stress failures.
2 The space between annealed glass and the shade should be a minimum of 2 inch (49 mm) or the space should be vented.
3 The glazing pocket must allow for the thermal expansion of the glass.
4 A heat source between the glass and the shade can build up heat on the glass.

Heat Register Location: Heating elements for the building should be located on the interior side of draperies. These elements should not be located between the shade device (draperies) and annealed glass. Register vanes should direct air away from glass. Thermal shock to the glass must be avoided during the initial start-up and balancing of the heating ventilating and air conditioning equipment.

Shading Devices: Blinds located on the inside of a window can trap heat and greatly increase the temperature of the glass. Air space between glass and indoor shading allows for ventilation and helps to reduce edge stresses. Natural ventilation is also aided with openings at the head and sill of the shading device. As the space between the glass and the shade decreases, heat builds up and edge stress increases.

Multiple Glazing: The thermal stresses produced in double-glazed units are usually greater than those induced in the same glass when single-glazed. This is because the insulating effect of air space reduces the heat loss from the outer pane to the inside of the building. This problem is more pronounced in triple-glazed units, especially for the center pane. Some window and door units allow glass panes to pass in front of or behind one another. This overlapping position increases the potential thermal stress in the glass.

Outdoor Shading Patterns: Tension stress, induced by thermal loading, increases appreciably with partial outdoor shading. Adjacent buildings, balconies, awnings, framing members, solar screens, louvers, trees and shrubbery may introduce shading stress. Horizontal shading, diagonal shading and vertical shading are less severe than double diagonals shading which essentially create shadows on top of shadows. The slight temperature differential caused by double shading may not appear to be critical; however, every degree lower at the glass edge will contribute to thermal stress failure. For every 1° F of temperature difference between the center and edge, there is approximately 50 psi (345 kPa) of stress applied to the glass. For example, if the edge of glass is at 50° F (10° C) and the center is at 110° F (43° C), the 60° F temperature differential will cause a 3,000 psi (20,670 kPa) stress in the glass. This is enough force to break annealed glass.

Recesses: Deep recesses in large projections in the façade can put portions of the glass in shade. Typically, the shaded areas are at the perimeter of the glass, which is unfortunately the weakest area of the glass. This is due to imperfections (chips, flakes or pits) resulting from cutting the glass. With other areas unshaded, thermal stress can develop.

Framing Systems: Framing systems of low heat capacity tend to minimize edge stress. For example, structural lock strip gaskets make excellent glazing rabbets because they are black and insulating. Lock strip gasket frames tend to reduce thermal stresses in glass. On the other hand, massive concrete frames can increase the thermal stress in glass because they have a large heat capacity and will keep the glass edges much cooler than the center of the glass pane. Low temperatures at the edges are sustained by an easy path of conduction from the glass edge through the frame to a cold building structure with a high thermal inertia. Concrete becomes cold at night and does not warm up as fast as glass when the sun shines. Framing systems should be designed to make glazing easy. If glazing is difficult, the glass may be inadvertently damaged during installation.

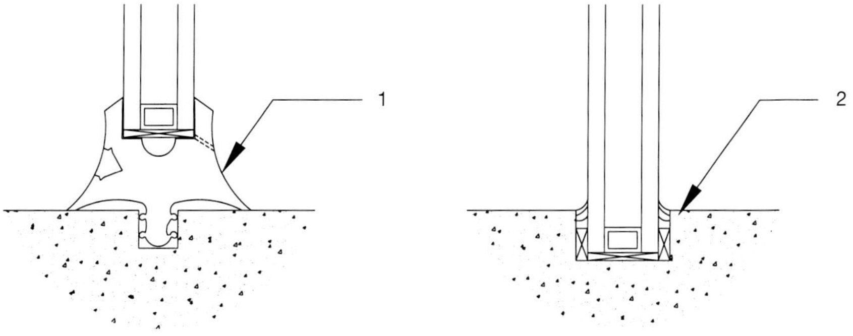

above Framing systems:
1 Gasketed framing systems of low heat capacity tend to minimize edge stress.
2 Concrete frames in contact with the glazing rabbets have a large heat capacity; the glass can thus experience high edge stresses.

below Outdoor shading patterns:
1 Outdoor double diagonal shade lines can contribute to the temperature differential between the edges and the center of the glass.

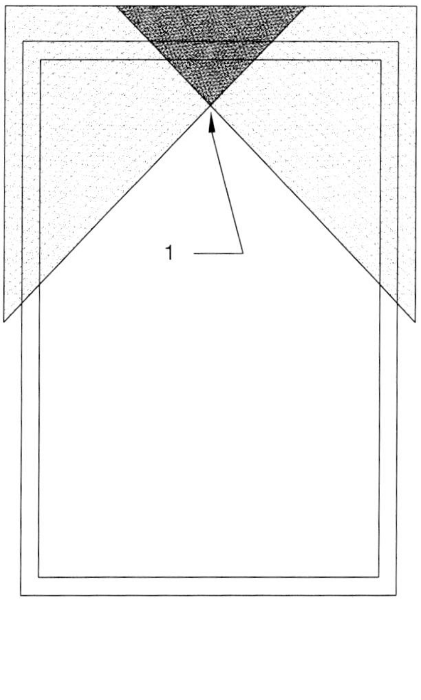

Size and Shape: Because thermal expansion is directly linked to the length of material affected by temperature, larger lights of glass are susceptible to greater thermal stress. Large lights with consequently large surface areas absorb more solar energy and may be exposed to greater edge stress. Irregular shaped glass requiring pattern cut shapes should be avoided if the glass is thicker than 1/4 inch (6 mm). Smooth cut lines on irregular shapes are more difficult to maintain with thicker glass fabrication.

Thickness: Thicker tinted glass absorbs more heat and therefore creates a greater range in temperatures. Thicker glass is more difficult to cut and therefore has a greater susceptibility to edge defects.

Tinted or Painted Glass: Used to reduce heat inside of our buildings, tinted glass absorbs more heat and increases the surface temperature of the glass. Thermal stress is minimal on thin, clear glass because the majority of solar energy is transmitted through the glass.

Reflective Coatings: Reflective coatings as well as the lower light transmitting "high performance" low emissivity (low e) coatings applied to tinted glass require special consideration. Certain types of body tinted glass and coatings on annealed glass are inherently at risk of thermal stress breakage simply because they add to the temperature imbalances which occur in the glass. The coatings absorb solar radiation in exposed areas versus the captured (covered) glass edges. The problem can be further exacerbated by reflective glass at the inside corner of a building, which will not only receive the heat from the sun directly, but also the indirect reflected rays from adjacent panels. Inside corners of buildings with reflective coatings can magnify the intensity of solar radiation on glass contributing to thermal stress.

Field-Applied Coatings and Films on Glass: Building owners occasionally want to modify the performance of their enclosure after it has been erected. The goal is to improve the envelopes' control of solar heat gain, thermal performance or visual comfort. One of the easiest ways to modify a glass enclosure is to apply a reflective or heat-absorbing film or coating to the inside surface of the glass. The most common materials used are polyester film and flow-on acrylic coatings. These materials alter the thermal characteristics of the glass. The film and coatings can be effective in improving the thermal comfort and reducing the effects of solar glare. These materials reflect a portion of the radiation; however, they absorb a high percentage of heat from the sun. The increased heat caused by the applied products increases the thermal stress within the glass light when exposed to the sun. The danger in applying films and coatings to existing annealed glass enclosures occurs if they are applied without evaluating the glass properties prior to application. If a heat-reflecting or absorbing coating of film is applied to the inboard surface of annealed glass, the increased thermal stress may cause the light of glass to break.

All of the above conditions contribute to thermal stress in the glass. Glass manufacturers can calculate the risk of thermal stress. But what can be done to prevent glass from breaking under these conditions?

above Installation of annealed glass painted black in lieu of heat-strengthened glass resulted in thermal stress cracks shortly after installation. Miami International Airport, Miami

page 35 Glass replacement has been required at the CNA Building, Chicago, due to thermal stress breakage.

How can Heat Stress Fracture be Avoided?

The traditional method for preventing glass failure due to thermal stress is to heat-strengthen the glass. Typical annealed glass windows with good edge fabrication may be expected to carry thermally induced stresses approaching 2,500 psi (17,235 kPa). Solar breakage risk can be reduced by heat-strengthening because the process increases edge strength by at least 2,000 psi (13,788 kPa). Heat-strengthened glass consists of annealed float glass that is cooled quickly, so that the outer surface cools faster than the core, resulting in residual stresses with the surfaces in compression and the core in tension. This effective prestressing makes the glass much stronger (twice as strong as annealed glass) and less susceptible to fracture due to thermal stress.

Another advantage to heat-strengthening the glass is weeding out the panels with edge defects. A large, deep chip is easily recognized on a glass edge; however, a tiny fissure is much less conspicuous. Minor defects to the glass will be uncovered during the heat-strengthening process because exposure of the glass to the great temperature differential involved with heat-treating will break the defective glass. Tempered glass is not typically susceptible to thermal stress failure for the same reasons.

If heat-strengthened glass is not used in locations where it is required, it will become painfully obvious shortly after installation. Miami International Airport was designed to have a black ceramic frit applied to the inside surface of the spandrel glass. The ceramic frit application to the glass involved a heat treatment that essentially heat-strengthens the glass in the process. In the airport project, the spandrel glass was inadvertently not coated with a ceramic frit. Instead, annealed glass, painted black on the inside surface, was installed. The black opaque paint absorbed the heat from the hot Florida sunshine. Thermal stress cracks quickly appeared shortly after the glass was installed.

Lessons Learned

1
Heat-strengthened glass greatly reduces the risk of thermal stress failure. Large lights of tinted glass should be heat-strengthened due to the thermal expansion of the material. Heat-strengthening can also reduce the need for thicker glass, an additional benefit in preventing thermal stress failure.

2
Glass edges are typically the weak point of glass, due to imperfections. Edges must be cut clean and protected. One must be careful not to bump or brush edges against metal or other hard surfaces. Quality control and tight fabrication specifications for rejection of glass with defects will greatly reduce the possibility of thermal stress failure in windows.

3
Glass should not be modified in the field. Do not nip edges or scarf corners. Glass edges must be protected to prevent defects.

4
Sufficient ventilation between interior shading devices and the glass must be provided to prevent the build-up of heat. Heating elements must be located and operated without causing extreme heat build-up on annealed glass.

5
Because of the risk associated with field-applied films and coatings and the thermal stress that comes with their application, these products should be used with caution on annealed glass.

6
Glass failure due to thermal stress can be identified through the close inspection of the cracked material. If the break line is perpendicular to both surfaces of the glass, the failure can be traced to thermal stress.

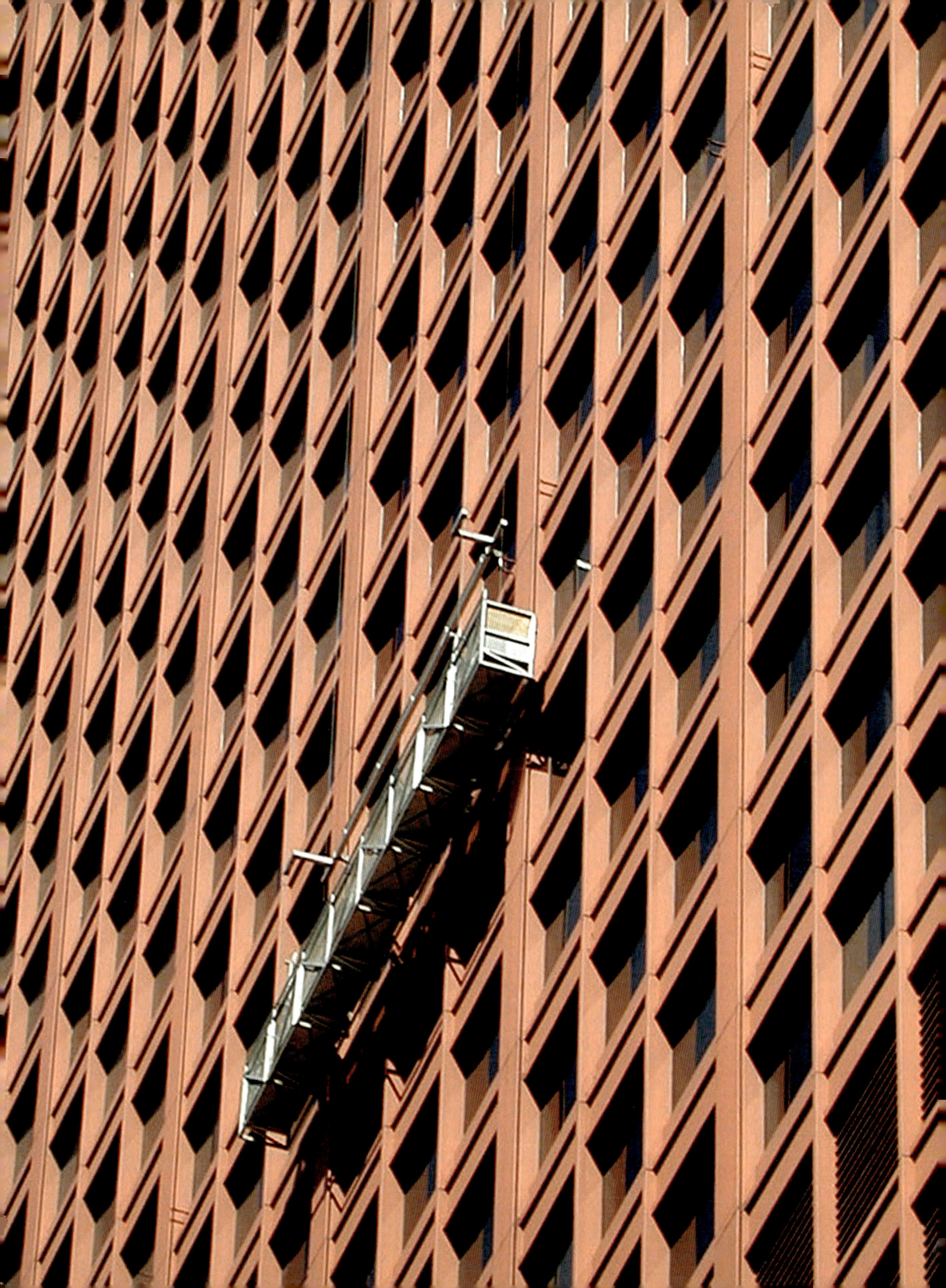

Corrosion

Weathering is a natural phenomenon. Its visible effect marks the passage of time on our bodies, our environment and our buildings. Some people believe that glass is not affected by time. Many of the early cathedrals provide proof of the durability of glass. Much of the 12th and 13th century glass is perfectly sound. The resilient, inert nature of glass is one of the reasons for its abundant use in our buildings. However, glass, like all building materials, can be affected by weather. To corrode means to eat or be eaten away gradually. Glass can corrode. When glass is exposed to moisture, a microscopic amount of the glass dissolves. The first interaction between glass and water results in the replacement of the sodium ions in the glass with hydrogen ions from the water. If the dissolved material is washed away, the human eye can see no change. Glass corrosion is inevitable whenever glass is exposed to water or other kinds of moisture. The longer the water stays in contact with the glass, the greater the amount of ion exchange. This initial corrosion is not discernible to the human eye, and as long as water is uniformly sprayed on the glass, no color difference will be visible.

If water is not uniformly applied to a glass surface (i.e. a slow drip from above), discoloration can occur. Chemists recognize two basic stages of glass corrosion due to water. The first stage occurs when sodium ions in the glass are replaced by the hydrogen ions in the water. This chemical exchange leads to a high sodium concentrate on the glass, which actually makes the glass more transparent. The second stage of glass corrosion, however, destroys the chemical bonds between the complex silicon, sodium, and oxygen tetrahedral network on the glass surface. When this structure falls apart, the visibility through the glass is impaired. White streaks and iridescent films can develop on glass as a result of a chemical reaction between the water solution and the glass. Alkaline substances such as concrete wash-off have been blamed for many of these problems; however, it is important to realize that they are not the only sources of this type of damage.

right Stain glass windows on old buildings can give the impression that glass does not change with time.
Parliament Building, London

page 37 Seagram Building, New York

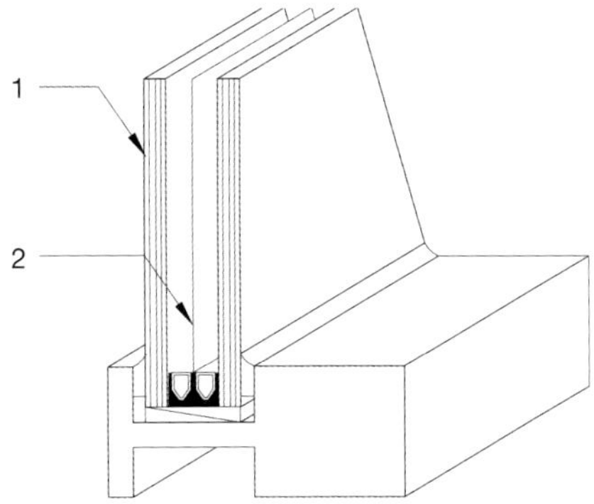

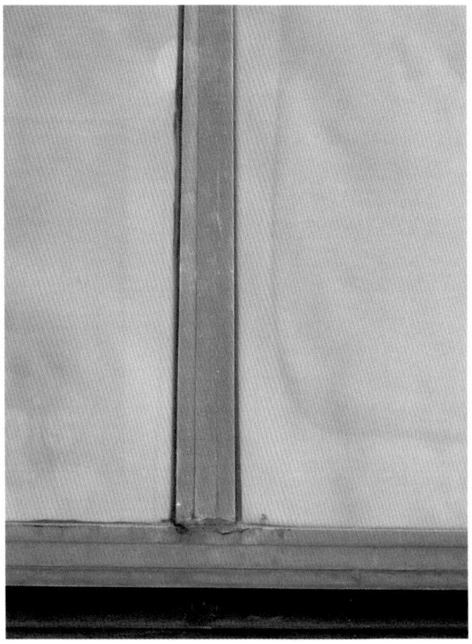

left Detail of insulating glass unit (1) with internal film (2). Early versions of film were coated with silver that can oxidize if exposed to moisture.

right This suspended metal-coated film inside an insulation glass unit displays corrosion.

below Completed in 1958, the dark amber tinted glass has displayed damage from etching.
Seagram Building, New York

Because of its visual function, glass is subject to a much higher level of inspection than other building materials. Staining is more noticeable on tinted glass because of the greater contrast between the light colors of the staining or etching and the darker color of the glass.
For this reason, surface damage that might remain unnoticed in other products like metal and stone may inspire serious criticism when this occurs on the surface of glass. The human eye is very sensitive to subtle variations in surface texture that can be magnified under certain lighting conditions. Pollutants and chemicals can accelerate the corrosion process. Severe examples can consist of pitting and eventual flaking of the glass. Water run-off from adjacent materials can damage glass with staining and etching. If the head conditions are not detailed to drip water away form the glass, rain run-off can permanently damage windows. Although 100 year old glass may show no visible detection of corrosion, one year old glass that has not been detailed properly or maintained adequately can exhibit pitting and flaking of the glass surface that is visible to the human eye.

The subtle change in glass exposed to weathering is tame as compared to the weathering of other materials. Metals can become patinated and rusty, thin stone panels can weather and lose strength while woods can rot and fade, and concrete can crack. Glass properties remain relatively unchanged with time and weather; however, surface damage of glass caused by a number of factors can reduce the strength and affect the appearance of glass. Surface damage of glass can occur from water run-off from adjacent materials. When water runs off masonry and some sealants, it can contain harmful contaminants that may react and adhere to the glass. Unlike corrosion, these contaminants can leave a residue on the glass that is difficult to remove. It is possible for the residue to stain or etch the glass, permanently damaging the surface. The corrosion of glass may be unattractive; however, corrosion of materials in contact with glass can cause the glass to break. This chapter will discuss how glass can corrode and how corrosion of other materials within an enclosure system can lead to glass failures.

right Completed in 1965, the Civic Center was the first government building constructed in Chicago in a non-traditional style.
Civic Center, Chicago

center The corrosion of the Cor-ten curtainwall frame exerts force on the glass which can cause breakage.
Civic Center, Chicago

below Details of typical windowsill, Civic Center, Chicago:
1 Tinted annealed glass
2 Cor-ten steel window frame
3 Thin glazing rabbet

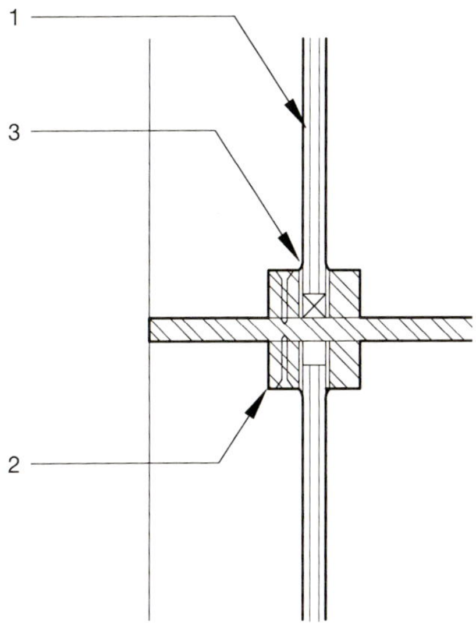

Completed in 1965, the Chicago Civic Center (later named the Richard J. Daley Center) was Chicago's first major public or government building to be built in a non-traditional style. The structure's rigid frame was, at the time of the building's construction, the largest all-welded building skeleton. The glass frames are composed of a special steel product manufactured by U.S. Steel. The curtainwall of the Civic Center is composed of Cor-ten steel frames holding 9 x 10 foot (2.74 x 3.05 m) tinted monolithic annealed glass panels, $1/2$ inch (12.7 mm) thick at the corner units and $3/8$ inch (10 mm) thick at the typical units. Cor-ten steel oxidizes with a self-weathering alloy containing a high carbon and potassium carbonate content. Provided that the material is not submerged in water, the Cor-ten steel's corrosion virtually stops at a certain stage, leaving the exterior surface without need for painting and maintenance-free.

The rustic brown color of the building has been a distinctive feature of the façade but has also contributed to a history of glass failure. In the first six years after construction, 200 of the 2,350 glass windows had broken. The large lights of glass were $1/2$ inch (12.7 mm) thick in some areas of the building and weighed as much as 635 lbs (287.7 kg) each. The natural corrosion of the Cor-ten frame has caused stress cracks in the annealed glass. A

right The curtainwall's brown color comes from the oxidation of the Cor-ten steel. Civic Center, Chicago

below Example of rusting steel storefront base which has been exposed to salts from winter de-icing. DuSable Hall, De Kalb, Illinois

thin sealed glazing rabbet was used to separate the glass from its Cor-ten steel frame. Rusting built up protrusions which penetrated the seal and contacted the tinted glass resulting in glass failures. The project had a history of glass crack failures, due primarily to the natural corrosion of the steel frame. Compounding the problem was Chicago's air pollution. Air laden with sulfur and nitrogen oxides combined with particulate matter left a stain-like film on the glass. The staining could only be removed with a hydrochloric acid solution that contributed to the steel corrosion. Run-off from the hydrochloric acid contacting the steel frame perpetuated more Cor-ten rusting and more stress on the glass. In order to reduce the risk of fallout, a film was added to the inside surface of the glass. Other Cor-Ten buildings in the Chicago area avoided this type of glass problem through the use of neoprene gaskets and painting of the glazing rabbets. The Civic Center glass problems show that corrosion of the metal frames that hold the glass can contribute to glass failure. An understanding of how various metals weather will help protect designers from glass failure due to corrosion.

left The use of pre-treated bronze provides a new façade with a patina shell.
60 Queen Victoria, London

right The curtain-wall glass is hidden behind bronze brise-soleil grilles and perimeter struts.
60 Queen Victoria, London

Metal Corrosion

There are two general types of metals used in construction: ferrous and non-ferrous. Ferrous metals (for example, cast iron, steel) contain iron. Ferrous metals will corrode (rust) when exposed to weather. Corrosion is defined as the electrochemical deterioration of a metal caused by atmospheric or induced chemical reactions. A common adverse chemical that enclosure systems are exposed to is salt, since it is commonly used for removing ice around the enclosure. Stainless steel contains a minimum of 11.5 % chromium. While it is a ferrous metal it does have a very good resistance to corrosion. Nickel and sometimes molybdenum is added to improve corrosion resistance.

Non-ferrous metals or alloys (for example: aluminum, lead, zinc, chromium, monel, copper, bronze, brass) contain no iron and are less susceptible to corrosion; however, many will develop a patina. Patina describes the surface oxide, carbonate, sulfate, or chloride that develops on the surface of some metals. People are most familiar with this process through the example of old roof gutters. Installed with a rich copper color, the gutters become dark brown and later green in their evolution through the patina process.

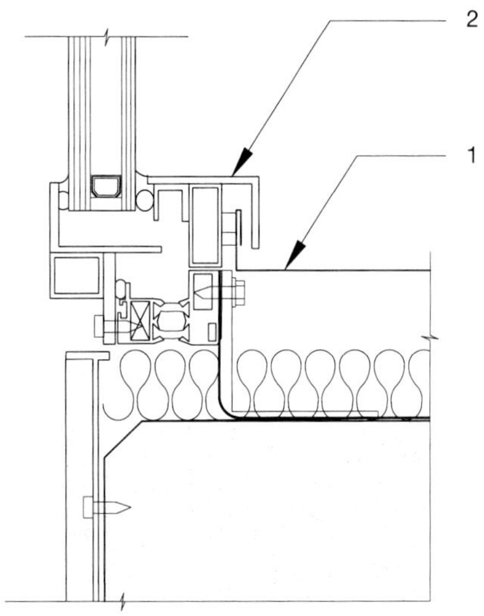

right Detail of windowsill indicates how glass is framed with stable aluminum; Jewish Museum, Berlin:
1 Zinc-coated metal cladding
2 Aluminum window frame

below A view of the window reveals that the metal encapsulating the glass is a non-weathering color-coated aluminum.
Jewish Museum, Berlin

page 43 The museum's walls appear like artwork from the interior as well as the exterior.
Jewish Museum, Berlin

The oxidation of metals used in enclosure systems does not necessarily lead to glass failures. Sometimes, the oxidation of cladding elements is expected, welcomed, and initiated by designers. Two projects that successfully demonstrate this are 60 Queen Victoria Street, London, and the Jewish Museum in Berlin. 60 Queen Victoria Street was completed in 2000. This brilliant work of architecture uses bronze façade elements pre-treated to accelerate the otherwise natural patina process. The eight-storey office building was clad with patinated bronze in order to have the building fit in with its traditional neighbors. The planners wanted a building with a solid feel, sympathetic with the highly articulated detail of surrounding stone buildings. The client wanted transparency. The architect's solution was to use patinated bronze: sheet, cast and extruded elements to create an enclosure that appeared solid, yet allowed light to pass through it. Bronze is easy and relatively economical to form, strong enough for castings – unlike copper – and can be patinated to give a varied, natural appearance. The result is a striking, modern façade, which looks solid but not heavy. Much of it is glazed, but because the glass is recessed behind bronze brise-soleil grilles and perimeter struts, it appears much less glazed than it actually is. The 12 foot 4 inch (3.75 m) window bays are set in bronze sheeting, interspersed with clear glass chevron windows. The patinating process was achieved in two stages: an initial application of a mix of acids for the dark base and a second, using blowtorch and stipple brush to bring out the green and achieve the patterned effect. The treated metal is then waxed. It is not waterproof, and the bronze will continue to oxidize on the exterior of the building. This artificial patina was the result of a done by hand texturing process and required special care to maintain consistency. The Queen Victoria Street building, which has cladding spread over 70,000 square foot (6,500 square meter), is estimated to be the largest expanse of patinated bronze in Europe at the time of its completion.

A second contemporary façade with changing metal is the Jewish Museum in Berlin. Completed in 1998 as an addition to a two-storey court of appeals building built in 1735, the museum is connected to its entrance by an underground tunnel. Uniquely separated from its Baroque partner, no passerby would confuse the Jewish Museum with an office building in Berlin. The glazing system reads like artwork from the interior and exterior. The unique window openings slash through the exterior surface of the building, dramatically offering light into the gallery spaces. The exterior skin of the building is composed of zinc-coated metal. As the building ages, the zinc will become patinated, providing a wonderful color and texture to the façade without causing failure to the glass. Although the building enclosure is made of a self-healing zinc carbonate material, the glass is captured in a coated aluminum frame.

Both the Queen Victoria Street building and the Jewish Museum show how metal oxidation can successfully add to the character of our buildings without compromising the glazing system. There are, of course, projects where the oxidation of metal claddings was not expected nor appreciated. Completed in 1998, an administration building added to the Aéroport Bordeaux Mérignac used aluminum panels to clad their façades. The panels were anodized to prevent oxidation and provide a uniform silver color. It is important to remember that raw aluminum does discolor if exposed to the atmosphere. When aluminum oxidizes – unlike steel – it turns chalky white. In the case of the airport panels, installation required cutting of the anodized material prior to installation. This left the cut edge of unfinished aluminum exposed. The result was a picture frame effect of discolored aluminum around the installed panels. The discoloration of materials due to weather is not limited to metal framing elements. A similar type of discoloration can occur in metal coatings applied to glass to improve their thermal performance.

Corrosion of Glass Coatings

The thermal performance of glass can be greatly enhanced by adding transparent thin metal films to one or more glass surfaces of an enclosure. The metal film enhances the thermal properties of the glass; however, it is susceptible to corrosion if unprotected. Glass coatings used in commercial buildings are available in two forms: Pyrolytic Coatings and Vacuum Deposition Coatings (Sputter).

Pyrolytic Coatings are thin metal oxide coatings applied to glass during the glass manufacturing process, and are, therefore, only available from a float glass producer. Vacuum Deposition Coatings are thin metal films applied to the glass surface after the glass manufacturing process. In this process, an ionic bombardment of a metal target plate with gas atoms dislodges ions of metal from target materials, depositing them onto the glass surface. The highly uniform transfer of metal, atom by atom, to the glass surface results in a very consistent product. Both Pyrolytic and Vacuum Deposition Coatings are used in reflective and low-emissivity glass.

Solar Reflective Coatings derive their name from their ability to reduce solar heat gain by way of higher reflection and absorption. The heat gain inside a building can be reduced dramatically by using these products but the trade-off will be lower light transmission through the glass. Low-Emissivity Coatings allow short-wave energy to be transmitted through glass, but reflect long-wave infrared energy. Both coatings improve the thermal performance of the glass by allowing less heat to transmit through the glass. Because the coatings contain metal, they are not stable in an open environment and are susceptible to change if exposed to moisture. Typically, the coated surfaces are protected inside insulated units; however, if they are exposed to wetting due to failed hermetic seals, they can corrode.

Corrosion of Fasteners

Corrosion of fasteners can lead to staining and ultimately curtainwall failures. Unexpected leakage into the enclosure system can cause corrosion of fasteners that hold the system together or connect it back to the building structure. Fasteners exposed to possible wetting must either be made of non-corrosive metals – for example, stainless steel – or coated with a protective material, such as obtained by galvanizing. Unprotected carbon steel fasteners used in curtainwall systems can cause long-term problems like staining of adjacent materials with rust marks and compromising of the structural integrity of the system.

page 44 Reflective coatings are susceptible to corrosion if exposed to moisture.

right An administration building is clad with anodized aluminum panels.
Aéroport Bordeaux Mérignac

below The anodized panels have started to discolor at the perimeter due to corrosion of the untreated cut edges.
Aéroport Bordeaux Mérignac

Etching and Staining

When water reaches a building, it is either reflected, absorbed, or allowed to run down the building's façade. When water run-off occurs across a masonry or concrete wall, the water can carry minerals from these surfaces to the glass below. Water run-off containing contaminants can damage glass. To understand this process, one must review some basic chemistry concepts. An aqueous solution can be described on a pH scale from one to 13, with neutral solutions having a pH of seven. Liquids measured below seven are considered acids, and liquids above seven are basic or alkaline. If the alkalinity of a solution reaches a pH greater than nine, the second destructive stage of glass corrosion can begin to occur. Fresh concrete can have a pH of 12; however, as the concrete ages, the alkalinity of the surface decreases. Damage to the surface of glass can be caused by alkalis that leach out of precast concrete panels due to rain or the fluorides in the wash-off from concrete floors that have been treated for hardening with solutions of zinc or magnesium fluosili-

right Dirt accumulation on glass can cause permanent staining.

below Circuit Court Building, Skokie, Illinois

page 47 White streaks on glass from water run-off can be difficult to remove.

cates. These substances can stain or etch the surface of the glass if allowed to remain for an extended period of time.

The contaminating effects from run-off water may lead to white streaks on the glass which are difficult to remove. The transparency of glass can be permanently lost once the glass surface has been severely etched. Although concrete and stone have taken most of the blame for unsightly glass stains and etching, it is important to remember that many glass structures built of materials other than concrete and masonry have experienced glass staining. This staining can be explained by the natural corrosion of glass. When uniform wetting of glazing occurs, visible staining of glass is generally not noticeable. However, when differential wetting of the windows occurs from a slow run-off of rainwater, such as dripping, a visible stain can occur regardless of the construction material used above. Any factor which increases the length of time moisture is in contact with glass during wet-dry cycles is likely to speed staining. One of the most common contributors of differential wetting is dirt. Dirt accumulation on glass holds the water on the glass longer, causing moisture to attack the surface. With time, a silica gel builds up and becomes hard. The hardened material retains moisture and guides run-off water along the same paths. The process of surface corrosion then becomes self-perpetuating. Frequent washing of windows tends to remove the dirt and gel before it becomes hard, thus minimizing the risk of staining and etching.

Chemical Attack

Intense staining normally implies a chemical change or degradation of the glass surface. However, flat glass used in construction is very resistant to chemical attack. Phosphoric and hydrofluoric acid are among the few chemicals that are harmful. Either of these substances in concentrated form will rapidly etch a glass surface and is commonly used for this purpose. Acid etching is a popular technique used to provide desired patterns in glass. Hydrofluoric acid is also used for cleaning new concrete. It is important to protect the glass from the hydrofluoric acid during the cleaning process. If unprotected, staining of the glass can become severe very quickly. Some alkaline substances will also attack glass surfaces to varying degrees. In concentrated forms, etching of glass surfaces can occur very quickly, within hours of contact. Diluted alkaline solutions are often used as glass cleaners and attack the glass surfaces extremely slowly. Glass surfaces must be thoroughly rinsed after cleaning to avoid staining. If not, the remaining residue from a concentrated cleaning solution can cause serious etching in a few days. Concrete surface treatments like sandblasting, acid washing, and waterproofing should be completed prior to glazing. Alkaline materials such as concrete wash-off and certain cleaning agents may chemically attack the glass surface. Silicone sealant, which inadvertently overlaps the joints onto the glass, can leave a mark. Because the silicone is so durable, the mark can become permanent on the glass surface. Careful masking of the sealant joint during installation can prevent this type of discoloration. Glass should be examined during the construction process to confirm that residual materials are not remaining on the glass. If these materials are detected, they should be rinsed off to prevent future damage.

Another common cause of staining is the use of lubricants for gasket-glazed installations. The mounting of a preformed gasket onto the enclosure frame is greatly facilitated by the use of lubricants. In some instances, the lubricants used have a high alkaline content (concentrated liquid cleaners) and cause staining of glass from rain run-off. In this case, even though the glass is rinsed, some cleaner is trapped between the gasket and the glass. This cleaner will seep out onto the glass over a period of time, and a serious attack of the glass surface may result. Only lubricants recommended by the gasket manufacturer should be used.

right The stains from the patina of metals can on occasion contribute to the design of an architectural element.

below Typically, the corrosion of carbon steel leads to unwanted staining of other materials below.

Professional glass cleaners often apply acid washes to the glass to remove the mineral deposits. This process can cause the surface of glass to appear milky white. This unwanted side effect is called acid burn. In addition, the acids can accelerate the corrosion of the aluminum mullion and deteriorate the sealants. Many washing mixtures contain hydrofluoric acid. If the acid content is high (greater than 6 % by volume), it can immediately damage the glass when combined with the heating effects of the sun. The hydrofluoric acid will start to dissolve the glass surface and make it more vulnerable to scratches from cleaning workers. Glass should not be cleaned while exposed to direct sunlight. It should be cleaned by starting at the top of the building, and systematically working down the façade. This technique will reduce the possibility of residue and cleaning solution running down previously cleaned glass.

Welding Sparks

Welding sparks can cause damage to the glass on a building. Because the temperature of molten metal is higher than the melting point of glass, the sparks can pit the surface of the glass. In addition to the visual imperfections caused by weld sparks, the small brown specks caused by pitting reduce the strength of the glass. This reduced strength can lead to glass breakage.

left Windows with failed hermetic seals can accumulate condensation regularly which may then cause wet-dry staining.

right Iridescence is an oil-like film usually caused by wet-dry staining.

Wet-Dry Staining

It is possible for the glass surface to become stained when condensation forms on glass in a repetitive manner, drying after each cycle. Moisture, in a thin film, can dissolve alkaline materials (like sodium) from glass into the water. If the water is a thin layer on the glass, as with condensation, the surface area of the glass is large in comparison to the volume of water. As sodium accumulates in the film of water, the liquid will increase in alkalinity at a much greater rate than if it were absorbed into a large volume of water. Many researchers agree that with high temperatures and no ventilation, as the alkalinity of the solution reaches a pH greater than nine, the second destructive stage of glass corrosion will begin to occur. If the corrosion is uniform, it will take on the appearance of a light fog on the glass that may go unnoticed to the naked eye. If condensation accumulates and covers an irregular surface, like dirty glass, the staining may become more noticeable as white streaks. Over extended periods of time, the glass surface in contact with the water will be removed and simple cleaning will not restore the transparency of the window. A common example of this problem is the discoloration of glass from failed insulating glass unit seals. Moisture is normally prevented from accumulating between the lights of glass of an insulating unit by placing desiccant inside the aluminum spacer. Preventing moisture from entering the air space is the task of the perimeter hermetic seals. Windows with failed hermetic seals can accumulate condensation on a regular basis. Condensation forming between the lights of glass can often lead to wet-dry staining.

Iridescence

Iridescence is an oil-like film usually caused by wet-dry staining. An iridescent film can develop on a glass surface if it is exposed to direct contact with water for extended periods of time. It can be very difficult to remove this discoloration. If glass is stored in a location that is exposed to moisture, water can be trapped between two lights of glass. The trapped water can leave an iridescent film on the glass that is permanent. Although iridescence can be removed, glass replacement is often the easier solution. It is important to

Acrylic glazing can discolor with time.
Munich Olympic Stadium, Munich

below Detail of acrylic glazing connection, Munich Olympic Stadium, Munich
1 Acrylic sheet
2 Stainless steel cable support system

page 51 Replacement of the acrylic glazing restores the original transparency of the enclosure.
Munich Olympic Stadium, Munich

remember that replacement of the glass can only be a partial and short-term solution to the problem. Determining the cause of the iridescent stain and preventing it from reoccurring is essential to keeping the glass transparent.

Iron Oxides

Glass may be damaged, etched or stained by a number of materials used at a construction site. Weathering steel, for example, releases oxides that can collect on glass. Rust (iron oxide) will not usually cause deterioration in the glass surface but may be very difficult to remove. These oxide deposits should be removed from glass by regular cleaning methods during the construction process. If the metal oxide deposits are permitted to wash over glass and accumulate, they can leave a tenacious residue. Landscaping sprinklers can contribute to glass staining with mineral deposits. Water from sprinklers often comes in contact with glass. As the water evaporates, mineral deposits can accumulate and adhere to the glass surface. Location of water sprinklers in relation to glass surfaces should hence be considered as a source of glass discoloration. The hard water from many sprinkler systems is rich in calcium carbonate. The water spray exposed to direct sunlight can produce tenaciously bonded inorganic residues on glass surfaces as the deposits are "baked" onto the glass.

Storage

Glass storage prior to installation can be a critical factor in the cause of glass staining. Stacked glass exposed to moisture and high temperatures is susceptible to staining. Lights of glass stored in an unventilated basement can result in the accumulation of condensation for long periods of time. In laboratory experiments, panes of glass stacked together in an enclosure that is heated and humidified continually for one week will become frosted and stained beyond reasonable repair. It is important to store glass in a place that is dry and well-ventilated. These precautions will virtually eliminate the possibility of such staining.

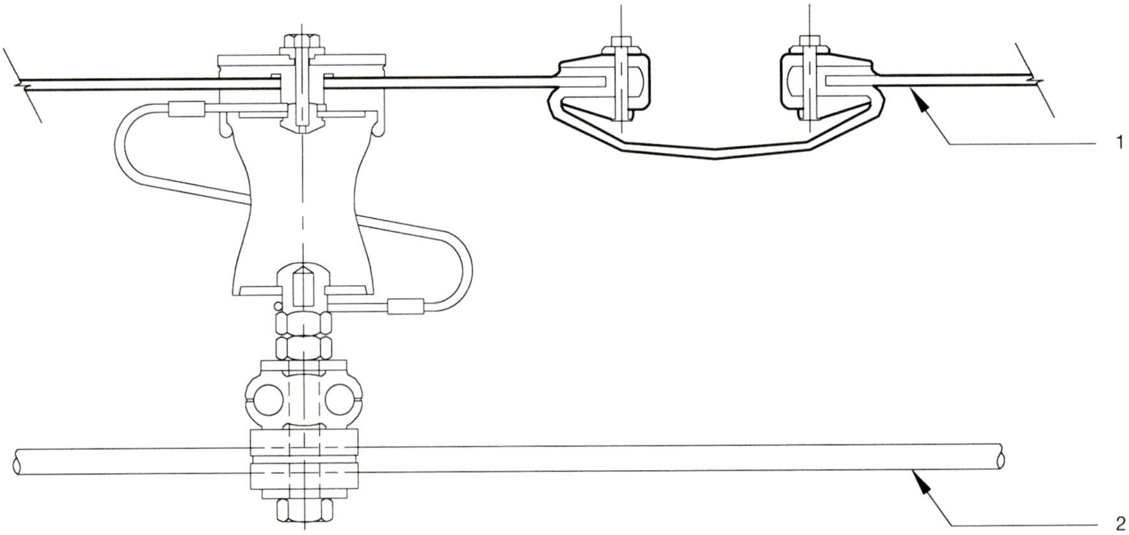

Discoloration of Plastic Glazing

Glass is not the only transparent enclosure material that can suffer from the effects of weathering. Plastic glazing is even more susceptible to the environment than glass. It is inherent in the material properties of plastic glazing that it becomes brittle and discolored with time. There are two types of plastic glazing products used on building enclosures: polycarbonate sheets (trade name Lexan by General Electric), and acrylic sheets (trade name Plexiglas produced by Rohm and Haas Company). Both products are combustible and cannot be used in fire resistance-related assemblies. When compared to clear glass of equal size and thickness, both products maintain greater resistance to impact and breakage and are lighter in weight. This makes them suitable for skylights and windows in vandal-prone areas such as correctional facilities and schools. They are bendable, have better acoustical characteristics than glass, and can be laminated to provide lightweight bullet-resistant barriers. Polycarbonate products are much stronger than acrylic. Although most materials are typically used in clear and transparent applications, polycarbonate sheets are available in light and dark bronze, opal and white. Acrylic products offer many color choices, including mirror coatings for interior applications.

Although plastic glazing is commonly used in windows and skylight applications, it is important to realize that their material properties are different from those of glass. Both polycarbonates and acrylics are more susceptible to scratching and abrasions than glass. Certain conditions of varying temperature and/or humidity on opposing surfaces of a single light may cause plastic panes to bow in the direction of the higher temperature and/or humidity. Though this does not affect visibility, it may

Head details above windows:
1 Provide a drip or gutter to keep water run-off from contacting glass

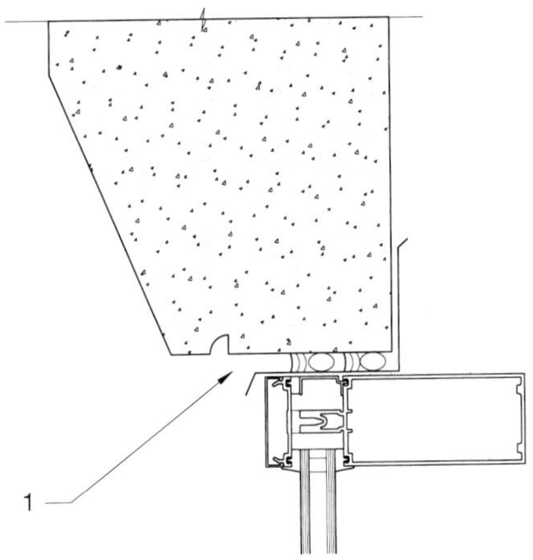

cause distorted reflections. Plastic products have a greater coefficient of thermal expansion than clear glass (approximately seven times greater than glass) and therefore are susceptible to higher degrees of dimensional change and should always have flexible installation connections that must allow for wide ranges of thermal expansion.

One of the largest and most innovative examples of plastic glazing is the Munich Olympic enclosure completed in 1971. The cladding for the roof was the subject of many debates during its inception. Alternate solutions ranging from lightweight sprayed concrete ("gunite") to a shell-type timber cladding were discussed during the design phase of the project. Eventually, the debate was resolved when the people responsible for televising the Games insisted that the roof be transparent. The plexiglas panels are attached to a cable structure providing the innovative solution to the request for transparency. The various Munich roofs cover 75,000 square meter (800,000 square foot). Individual cable nets with constant mesh width of 75 x 75 cm (2 foot 2 inch x 2 foot 2 inch) are surrounded by edge cables in order to carry their loads to singular supports. The enclosure is made up of a cable net mesh of locked coil ropes, cast steel connections, and masts made of steel tubes.

The acrylic glass covers have changed with time. They have been discoloring to the point that they are now only translucent. Recladding of the project was started in 1999. Nonetheless, there were no signs of breakage. This project does not represent an enclosure failure. The atmosphere affects almost all materials, and their changes should be anticipated. Unlike glass, plastic products change more with their environment.

How can Unwanted Corrosion be Avoided?

Most materials change with time if exposed to moisture. By understanding how materials change, one can work to protect them and prevent unwanted corrosion. Contemporary curtainwall façades encapsulate the glass in aluminum frames that are protected from corrosion by high-performance coatings. Coatings vary from paint to anodizing, to PVDF (PolyVinyleDene Fluoride, brand name Kynar), to powder coating. The common theme of successful façade design is the encapsulation of the glazing in a material that will not change when exposed to weather.

An effective method of reducing the staining and etching of glass surfaces is to use a drip edge above the window to direct run-off water away from the glass. This will prevent many staining and etching problems. Soffits and exterior window head details must be designed to direct water away from the glass in order to prevent damage. The use of reveals and drips can minimize the risk of run-off and help avoid the deposit of residue onto the glass.

"An ounce of prevention is worth a pound of cure." Run-off can contain chemicals from building materials above as well as pollutants from the atmosphere. Although dirt on glass will not always cause staining, proper cleaning of the glass will virtually guarantee that staining will not occur. In this sense, periodic window washing is critical to preventing stains and corrosion from occurring on glass. Glass enclosures should be cleaned from top to bottom. By systematically working down the building, the water run-off from the cleaning will not further contaminate the glass surfaces on the enclosure. It is not recommended to clean glass in direct sunlight.

Newly developed hydrophobic coatings can be applied to the glass surface to protect it from chemical attack. Hydrophobic coatings create a barrier between the glass and water and other pollutants. They help prevent the initial stages of corrosion and reduce the occurrence of other materials like mineral deposits from sticking to the glass. Hydrophobic substances repel water and harmful chemicals and keep glass free from staining. If the microscopic level glass surfaces are truly jagged with hills and valleys, hydrophobic coatings fill the voids creating a smooth seal over the glass surface. This smoother surface is more scratch-resistant, repels water, and provides less adhesion for unwanted materials. The performance of hydrophobic coating products can vary from weeks to over five years; however, they can be reapplied.

Lessons Learned

1
Glass is a stable material; however, it can be etched and stained by water run-off. Ledges and drip moldings will prevent run-off from contacting the glass surface and will minimize the risk of etching and staining the glass.

2
Early glazing frames were made of steel. These systems had glass breakage problems when the glazing rabbets did not provide sufficient space and cushioning material to allow for rust accumulation. Using neoprene glazing gaskets and painting the steel glazing rabbets reduces this risk. Most contemporary glazing systems use non-corrosive metals in close contact with glazing material.

3
It is important to understand how materials adjacent to glass change when exposed to weather. If oxidation is anticipated, separation of the metal from the glass elements is recommended. Protective coatings on metals exposed to weather can limit their susceptibility to corrosion; however, cutting of these elements in the field may leave unprotected surfaces exposed to the elements. This should be avoided.

4
Some material changes are expected, even desired in our building envelopes. The color and texture of patina metal add to their character. Nonetheless, the glazing details that relate to changing material must anticipate the change in order to prevent glass failure.

5
The cleaning of glass at the conclusion of the construction process is critical to removing harmful deposits on the glass surface. If residual materials are not removed, these materials can permanently damage the glass. Regular cleaning of glass during the life of a building is the best method of avoiding surface corrosion, staining, and etching.

6
Although similar to glass in transparency, plastic glazing material properties are very different from glass. Some of these properties such as strength are desirable, while some of the properties such as scratch resistance and discoloration are not desirable.

Incompatibility

When we say two people are incompatible, we mean that they are adversely affected by the other's presence. When in contact, they argue, fight and/or negatively influence each other. Incompatibility between building materials is similar. Like people, two materials can be stable when isolated; however, if brought in contact with each other, they can corrode, discolor, and lose adhesion. For example, aluminum in contact with wet concrete will decompose. Zinc can corrode if in contact with copper. Silicone will discolor if exposed to many asphalt substances. The transition from one material to another is critical to architectural glass detailing. A fundamental requirement for a successful transition is that the materials in contact with each other are compatible. This means that they will not adversely affect each other's performance after installation. Glass is by nature neutral with respect to its surroundings, yet some glazing failures are due to incompatibility of materials that surround the glass. Incompatible sealants and adhesives can discolor or lose adhesion if in contact with each other. Incompatibility can be chemical, leading to sealant breakdown, or physical, leading to the loss of adhesion. The plasticizer in some extruded gaskets can migrate to the sealant interface and cause a loss of adhesion shortly after installation. The affects of incompatibility can be as simple as not adhering. For example, urethane sealant will not adhere to silicone sealant. Incompatibility between materials can often be propagated by a catalyst like water.

Dissimilar metals can corrode when exposed to moisture or chemicals due to galvanic action. If enclosure materials are incompatible or become incompatible after installation as a result of wetting, their properties will be adversely altered. Sometimes the incompatible materials are not part of the original con-

above Aluminum in contact with wet concrete or masonry will decompose.

right The discoloration and partial delamination is only visible upon close inspection. Grand Louvre Pyramid, Paris

page 55 Using extra-clear white glass, the pyramid was the first structural silicone-glazed building in France. Grand Louvre Pyramid, Paris

above The sloped skylight provides an elegant entrance to the old museum.
Grand Louvre Pyramid, Paris

below left Detail of structural silicone glazing with laminated glass indicates how interlayer can come in contact with sealant and moisture.
1 Laminated glass
2 Silicone sealant
Grand Louvre Pyramid, Paris

below right Edge blushing: when the PVB interlayer of laminated glass comes in contact with silicone, it is common for minor delaminating (1/2 inch or 13 mm) to occur at the edges.

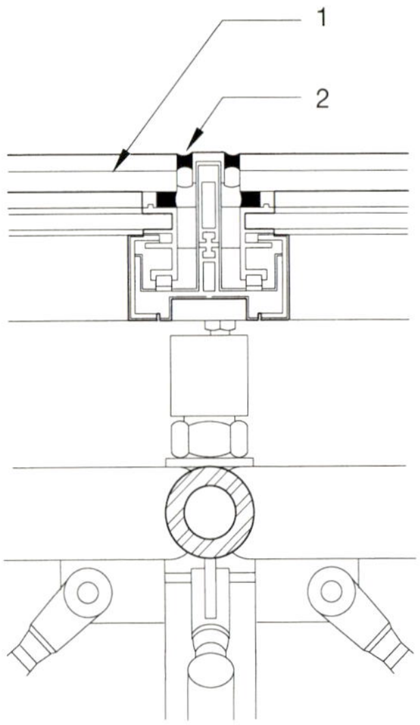

struction but later additions to the enclosure. One must have a thorough understanding of material properties in order to prevent compatibility problems. If not, material performance and or appearance can be affected. A minor defect that can be attributed to incompatibility can be seen at the Grand Louvre Pyramid in Paris. Completed in 1988, the project had a quest for transparency in designing an addition to one of the city's most treasured landmarks. A light filigree structure composed of stainless steel cables and connections support the resulting skylight system. A Massachusetts firm that produces riggings for the America's Cup yachts manufactured most of the connections. The light elements were designed to reduce the visible mass of the pyramid structure. The glazing is composed of 3 x 1.9 m (8 foot 8 inch x 5 foot 6 inch) rhombus-shaped panes. Most float glass products contain small amounts of iron oxide, which produce a green tint to the glass. The green tint is most noticeable by viewing the glass edge. In order to avoid the greenish tint of iron float glass, a laminated plate glass system with a negligible iron oxide mix was used. The glass specified was an extra-clear white glass previously only fabricated for use in solar-collector mirrors. At the time the Pyramid was built, the industrial manufacturing processes were not capable of producing this type of glass in the sizes desired, so the panes had to be poured, ground and polished by hand in the UK. The polished glass is bonded to an aluminum skylight frame by means of silicone sealant. This was the first installation of structural silicone glazing completed in France. The narrow joint between the glass and the frame is filled with a silver-colored silicone sealant. Unlike traditional glazing systems, four-sided structural glazing has no aluminum caps to cover the joint between panes of glass. Some of the silicone from the exposed joint has come in contact with the PVB (polyvinyl butyral) interlayer. The resulting yellow discoloration and partial delamination of the glazing edge could be the result of material incompatibility. This does not represent a compromise to the structural integrity of the glazing system. It only represents a minor blemish to an otherwise pristine enclosure. Laminated interlayers have historically experienced compatibility problems with sealants. The formulation of silicone and interlayer has developed since this project was completed. Some compatibility concerns have been removed with certain products; however, verification through compatibility testing should be completed prior to glass fabrication.

right Laminated glass canopy with exposed glass edge
CNIT Canopy, Paris

center The canopy provides shelter to the entrance without blocking all of the sun's light.
CNIT Canopy, Paris

below Discoloration and delamination of exposed glass edge due to moisture.
CNIT Canopy, Paris

Interlayer contamination is not limited to silicone. The edges of laminated glass can be susceptible to delaminating if exposed to moisture. A glass canopy sheltering the new entrance to the CNIT building in Paris completed in 1991 shows signs of discoloration and delamination. The laminated glass surface extends beyond its steel truss support with an unprotected laminated edge exposed to view and weather. With time and exposure to moisture, the laminated edge is failing. Delaminating is extenuated by the discoloration that accompanies it. Laminated glass using poured-in-place resins has a better track record with exposed edges than PVB interlayers. A metal channel is often used to protect the laminate interlayer from incompatible silicone and water.

There are several types of incompatibility dangers that affect enclosure design. This chapter will review four: galvanic action of metal, the chemical reaction of materials, staining of glass, and the loss of adhesion of sealants.

Galvanic Action

The deterioration of one material caused by the contact with another material is a form of incompatibility. Galvanic action, a form of corrosion, occurs when dissimilar metals are in contact with each other and when sufficient moisture is present to carry an electrical current. The galvanic series describes a list of metals arranged in order from "most reactive to corrosion" to "least reactive to corrosion".

right The edges of laminated glass are often covered to prevent moisture from coming in contact with the PVB interlayer.
Illionois Center Canopy, Chicago

below Diagram of the Galvanic Series

Anode	Magnesium, Magnesium alloys
(least noble)	Zinc
+	Aluminum 1100
	Cadmium
	Aluminum 2024-T4
	Steel or iron, Cast iron
	Chromium iron (active)
	Ni-Resist
	Type 304,316 stainless (active)
Electric current	Hastelloy "B"
flows from positive (+)	Lead, Tin
	Nickel (Inconel) (active)
	Hastelloy "B"
	Brasses, Copper, Bronze, Monel
	Silver solder
	Nickel (Inconel) (passive)
	Chromium iron (passive)
to negative (-)	Type 304, 316 stainless (passive)
Cathode	Silver
(most noble)	Titanium
	Graphite, Gold, Platinum

The less noble a material is, the more reactive to corrosion it will be. The further apart two metals are on the list, the greater the polarity that will develop and consequently the greater the decay of the lesser noble metal. Separation of dissimilar metals with a neutral nonconductive material is fundamental to preventing galvanic action.

Chemical Reaction

Metal deterioration also occurs when metals come in contact with chemically active materials. Typically, this process can be accelerated when moisture is present. For example, aluminum in direct contact with concrete or mortar will corrode. Steel in contact with certain types of treated wood will corrode. An example of a chemical reaction occurring on a glazing system occurred at the Burroughs Wellcome Research Headquarters in North Carolina. Completed in 1972, the building stretched expansively across a wooded hillside ridge in North Carolina. This large corporate headquarters building's profile echoes the inclines of the hills around the site. The progressive design has received numerous awards and has served as the backdrop of a Hollywood film. The sloped enclosure system for the project had problems shortly after its installation. The enclosure was plagued with a transition detail that had serious compatibility problems. The cement stucco that surrounded the window system contained magnesium oxychloride cement plaster. The sloped glass was isolated from the adjacent material by a zinc separator. When the stucco got wet from rain, the oxychloride combined with the water to make hydrochloric acid. When water is present and in contact with zinc, the magnesium oxychloride initiates an intense galvanic reaction that causes corrosion of the sacrificial zinc material. Ironically, the system had been tested in a mock-up prior to fabrication. The mock-up testing of the system used plywood in lieu of stucco. Because the stucco system was not being tested for water infiltration, it was not included in the mock-up. The testing was successful, and the compatibility issue with the stucco was not detected. However, within weeks of completing the enclosure

above The building's sloped enclosure system echoes the inclines of the hills around the site.
Burroughs Wellcome Headquarters, Research Triangle Park, North Carolina

right The testing mock-up did not incorporate the stucco coating which later proved to be incompatible with the zinc divider strips on the Burroughs Wellcome building.

on the building, the zinc border had corroded, allowing water to leak into the building. The entire enclosure system was ultimately replaced.

Staining

Organic sealants commonly used in building enclosure systems may exude, bleed, or leach solvents, oils and plasticizers during the life of a building. Many factors affect how and how fast the residue will continue to exude from the sealant, including the weather conditions and the type of sealant used. If the glass is frequently cleaned after installation, the residual deposits from the sealants will be removed and not cause any problems to the glass. When the residue from sealants is allowed to remain in contact with glass surfaces over an extended period of time, the residue may bond to the glass surface and be difficult to remove. The glass need not necessarily be in direct contact with the exuding sealant to be affected by it. If sealants are used in areas around the glass and residue from these sealants runs off the material, it may leave residual deposits on the glass. The cleaning of glass at the end of installation is an important step to preventing residual sealant deposits from staining the glass.

Loss of Adhesion

A potential side effect of incompatibility is the loss of adhesion. In most cases, this will result in failed seals that allow water to penetrate the system. In the case of four-sided structural glazing, loss of adhesion can lead to glass falling from the building. A four-sided structural glazing system generally relies solely on the adhesion of the sealant to connect the glass to the metal frame. If the sealant does not adhere to the metal or the glass, the glass panel can be sucked off the building during a windstorm.

Early building examples of structural silicone sealant glazing systems struggled to obtain adhesion to anodized aluminum frames. It was, however, possible to identify the problem with testing prior to fabrication. Sometimes adhesion was determined to be erratic, i.e. adhering in some cases and not in others. With a large office building in Hong Kong,

a major problem could be prevented by early detection of lack of adhesion in the glazing system. During testing, it became clear that the silicone was not adhering to the anodized aluminum. Luckily, the problem was detected at the mock-up stage and the faulty materials were not used on the actual building. The project revealed the importance of sealant adhesion testing prior to fabrication.

Early silicones were all acetoxy and more susceptible to adhesion problems than the silicone produced today. Acetoxy silicones would stain and lose adhesion if in contact with organic spacers (for example, EPDM/ethylene-

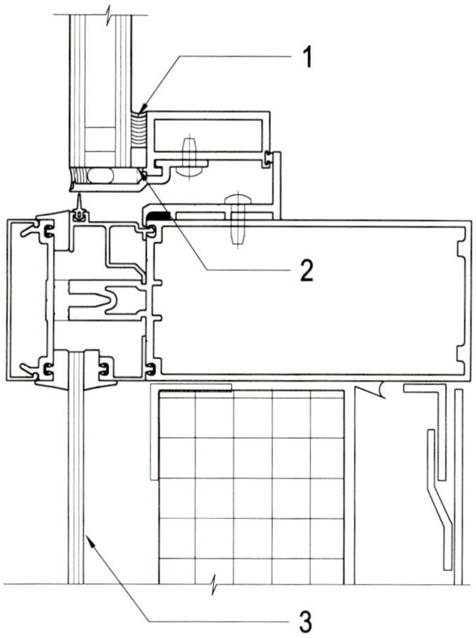

right Window vent detail with four-sided structural silicone sash
1 The insulating glass window is held on to the frame by nothing but the silicone sealant.
2 The heal gasket that comes in contact with the silicone must be compatible with the sealant.
3 The spandrel glass in this detail is supported by a standard pressure bar connection.

propylene-diene monomer, neoprene, sanoprene). The plastisizers used in the gasket extrusion process would migrate out of the gaskets and into the adjacent silicone sealant. The initial effect was staining of the silicone. Light-colored silicones would be discolored by the black marks of the leaching plastisizers. In addition to staining, the leaching reduced the adhesion capability of silicone. This was particularly problematic in the case of four-sided structural silicone glazing systems which rely on the adhesion of the silicone to keep the glass skin on the building. A loss of adhesion due to incompatibility of the silicone with an extruded spacer could bring about the tragic displacement of the glass. Early four-sided silicone project used acetoxy silicones that often struggled with adhesion problems. Although no glass had fallen off of the building, the entire façade of a 1960s office building in Miami, Florida, was reclad due to the loss of adhesion in an early developed four-sided structural silicone glazed wall. Today silicones are much stronger and sustain bond strengths as high as 80 psi (551 kPa). Proper curing of the silicon is important to maintaining these high bond strengths, and all gaskets in contact with silicone that rely on adhesion must be made of silicone or silicone-compatible material.

Opacifier Films and Coatings

When monolithic glass is employed as a spandrel, an opacifier is used to restrict viewing of the curtainwall insulation within the assembly. The best method for creating a uniform opaque background is a ceramic frit applied to the inside surface of the glass. Ceramic frits are manufactured through a heat process that certain glass products, for example some reflective coatings, will not withstand. With

below Loss of adhesion: early structural silicone sealant did not perform well under testing. The silver color of the anodized aluminum mullion indicates areas where adhesion was lost.

A second method of providing an opaque background on reflective glass is an opacicoat. An opacicoat is a pigmented silicone "paint" that is rolled or sprayed onto the inside surface of the glass. Like the opacifier films, opacicoats should not be in contact with building insulation. An example of an Opacicoat failure occurred at the 333 Wacker Drive project in Chicago. Completed in 1981, the building's curved west façade gently follows the line of the bending Chicago River. This award-winning project had a problem with the spandrel glass. The spandrel glass is the smaller rectangular light below the vision light, and it hides the transition space between floors. This is achieved with an opaque glass that has the same reflective properties as the vision light. The green-tinted monolithic spandrel glass had a coating on the inside surface that lost adhesion after installation. A possible cause of the failure of the opacicoat is that the spandrel insulation came in contact with the glass, contained moisture, and thus caused the peel-off. The problem was corrected in 1992 by replacing all of the spandrel glass.

Completed in 1979, the cathedral was one of the first large scale four-sided structural silicone sealant enclosures ever constructed.
Crystal Cathedral, Garden Grove, California

these types of reflective glass, a polyester opacifier film is used to cover the glass. The film is applied to the interior glass surface over the reflective coating.

Insulation must not be applied directly to the opacifier film. The areas behind the panel must be vented to prevent condensation from forming on the cold interior surface of the glass. Insulation should be installed with a 1 inch (25.4 mm) air space between the glass and the insulation. Metal retainer straps are often used to prevent the insulation from contacting the glass.

How can Incompatibility be Avoided?

Testing of materials prior to installation can help eliminate undesired compatibility problems. Common culprits of incompatibility are sealants; however, sealant manufacturers will gladly test their products for adhesion and compatibility to other materials prior to construction. Primers will often be recommended for application to certain substrates as a method of improving adhesion.

Separation of dissimilar metals through coatings will prevent galvanic action from occurring. Chemical reactions between materials can be avoided by researching materials prior to installation and verifying that exposure to the elements will not change their properties. Chemical reactions and galvanic action can occur with the passage of moisture from one material to another. For example, several meters may separate two metals from each other; however, droplets of water from one to the other can cause an ion exchange as if they were touching.

In addition to having a basic understanding of material properties, designers should thoroughly investigate new materials used in construction. Testing of material properties like adhesion must include all possible conditions to which the building is susceptible, including temperature changes, exposure to moisture, and ultraviolet radiation.

Lessons Learned

1
Testing of a system is crucial to the success of the enclosure system; however, it is necessary that the materials used in the testing are the same as those used on the project. Often substitute materials are employed, assuming that their role in the testing is inconsequential. This can result in compatibility failures with the actual materials.

2
Glass coatings can be adversely affected if in contact with insulation material. Insulation should be held off glass by a minimum of 1 inch (25.4 mm).

3
Solutions used to clean enclosure materials before and after installation must be reviewed to determine if they will have an adverse effect on the enclosure system.

4
Testing of materials for adhesion, staining, and compatibility is crucial to avoiding problems after installation. Sealants, gaskets, and interlayer materials should be tested for adhesion and compatibility prior to fabrication. The edges of laminated glass may require protection from moisture in order to minimize the risk of perimeter delamination.

5
Separation of dissimilar metals in the presence of moisture will avoid the risk of deteriorating the less noble metal due to galvanic action.

above The green glass of the tower follows the bend of the Chicago River. 333 Wacker Drive, Chicago

below and page 62 The spandrel panels were replaced following a loss of adhesion of the reflective coating to the glass. 333 Wacker Drive, Chicago

Leakage

In ancient times, when the weather turned foul, prehistoric man sought shelter in caves. As civilization advanced, man began to improve his shelters. Roofs went from opaque wood enclosures to all-glass surfaces designed to shed water. The evolution of our building skins has generated products that outperform the materials and techniques of the past. Yet with all of our advances, we still are faced with the age-old problem of leaks. It is still one of the most common problems with glass enclosure systems. The consequences of water leakage can be serious. Initial water damage to interior finishes and equipment can be followed by secondary effects like corrosion of anchors, rotting/mildew, material disintegration, loss of insulating value, and electrical problems. Water leakage can go undetected at its early stages and significant damage can occur. Air infiltration through a building skin can be a source of human discomfort, energy loss, and the accumulation of condensation.

Historically, people have tried to solve leakage problems by improving joinery. Improvements in sealants, gaskets and installation techniques have been effective in preventing initial leaks. Water leakage can generally be considered the result of three simultaneous conditions: an imperfect seal, positive differential air pressure (i.e. the air pressure outside being higher than the interior pressure) and the presence of water. When all three conditions are present, water running along a building surface is pushed through the imperfect seal by the pressure differential created by the positive winds. Recent techniques have shifted to eliminating the pressure differential at the surface of building skins. Pressure equalization has made great strides in reducing leakage into our buildings.

A common solution to water leakage in traditional enclosure systems is to find the source of the leak on the exterior of an enclosure and

right Roof enclosures have become far more complex; however, the need to keep water out has remained. Lucent Technologies, Naperville, Illinois

page 65 The Johnson Wax Administration Building in Racine, Wisconsin, was designed with an innovative translucent glass roof system to cover the great workroom. Although a breathtaking design, the original system leaked during rains since its completion in 1939.

repair it. Hardships arise when getting to the exterior is difficult, and when the problem repeats itself over large areas of the building enclosure. Regardless of this, a repair to the exterior joinery may only be a temporary solution. The outer barrier of many enclosures can never be homogenous like roof membranes. Masonry walls crack with time while panelized systems (metal, concrete, or glass) require gaskets and sealant joints at transitions that will eventually fail.

The most common cause of leakage into a building is sealant failure. External sealant joints by themselves do not provide a reliable and durable weather protection for an enclosure system. They are vulnerable to harsh weather conditions as well as to the field installation environment, which includes dirty substrates and sloppy workmanship. Relying on an outer barrier as the sole protection against water penetration for these types of systems can lead to water leakage problems.

Reliance on an exposed single barrier of sealant to prevent water penetration into a wall is the single most common source of current building façade problems. Despite substantial improvement in the weathering resistance of sealants, this material is not always reliable as the sole barrier against water and air infiltration. A single sealant barrier is only effective if the seal is perfect. But perfect seals require perfect substrates for the application and perfect workmanship. Over a large project, both of these expectations are unrealistic. A design that depends on such perfection is doomed to fail. The key to preventing leaks inside a building is to provide a secondary system of weepage. If this system is not effective, water will find its way into the building.

The Revenue Building in Springfield, Illinois, completed in 1983, was designed with a skylight system which covered a large percentage of the building plan. Large atrium spaces fill the interior of the building and are covered with 34,000 square foot (4,118 square meter) of laminated insulating glass. The skylight is supported by an open web tubular truss system. The two rectangular towers of the building are connected with a series of interior walkway bridges. From the bridges and balconies of the center atrium, one can look through the skylight to a breathtaking view of the domed state capital building to the south of the site. Unfortunately, these views have been tarnished by a skylight system riddled with problems.

page 66 Horizontal mullion caps prevent water from shedding off the skylight and contribute to leakage problems.
Revenue Building, Springfield

left Deterioration of skylight has tarnished views of the capital building.
Revenue Building, Springfield

below View of interior atrium and walkway bridges
Revenue Building, Springfield

Details of original skylight system, Revenue Building, Springfield:
1 Water cannot shed off horizontal mullion.
2 Condensation gutter and internal drainage cavity
3 Condensation gutter and internal drainage weep into small gutter.
4 Attachment to steel structure does not allow for erection tolerances.
5 Steel tube truss for support of skylight

Details of replacement system:
6 Horizontal mullion is structurally sealed, requiring no cap.
7 Condensation gutter
8 Collection gutter and internal drainage cavity are large and separate.
9 Connection to steel structure allows for erection tolerances.

From the time of its installation until its replacement, the skylight leaked to the interior of the building during rainy weather. Despite numerous attempts to correct the problem, water managed to elude contractors and maintenance workers and find its way to the floor below. The problems with the existing skylight could not be easily fixed. A detail of the typical skylight section reveals how water collected at the horizontal caps and submerged exterior sealant joints. This detail taxes not only the material properties of the sealant but the workmanship required to install the joints. The exterior pressure caps did not provide the continuity of seals required to prevent leakage into the system. Poor joint design linked with the thermal movement of the skylight caused the exterior skin to leak. The skylight's internal gutter was intended to collect incidental water from failed seals. It appears that the Revenue Building's skylight gutters were too small for the grand size of the project and the amount of water that was being collected. These tiny gutters could not accommodate the large volumes of water that infiltrated the failed exterior seals and transition details over the large expanse of skylight. The aluminum members of the enclosure system required splices between elements to accommodate the thermal movement of the material and the fabrication limitation of the extrusion process. The internal drainage system, which was intended to direct infiltrating water out of the building, did not properly pass water through these transition splices. In numerous locations, the water skipped over the splice, out of the gutter and ran down the steel pipe structure. Ultimately, the water from the pipes spilled down onto the floor of the building. Countless attempts to seal the joints from the exterior merely gave pause to the leakage of a system that had serious inherent problems. These temporary patches slowed the leaks to a less noticeable level; however, with time, the leaks returned more numerous than before.

Many of the lessons learned from the original construction were incorporated into the replacement system. The new skylight, installed in 2001, used a flush glazing system at the horizontal glass joints. Water was not able to collect at these horizontal mullions. A gutter system, which could accommodate water infiltration from the large area of skylight, was installed. Great care was taken to incorporate an internal drainage system that could bridge all transition details and carry penetrating water out of the building onto the adjacent roofs below. A mock-up of the skylight system received exhaustive laboratory testing prior to fabrication. Lastly, field water testing at four different stages of the project installation covering all major transition details was done to ensure that the skylight leakage problems of the Revenue Building were corrected. The building serves as an example of how skylight problems can be solved and building envelopes can successfully combat the forces of heavy wind and rain.

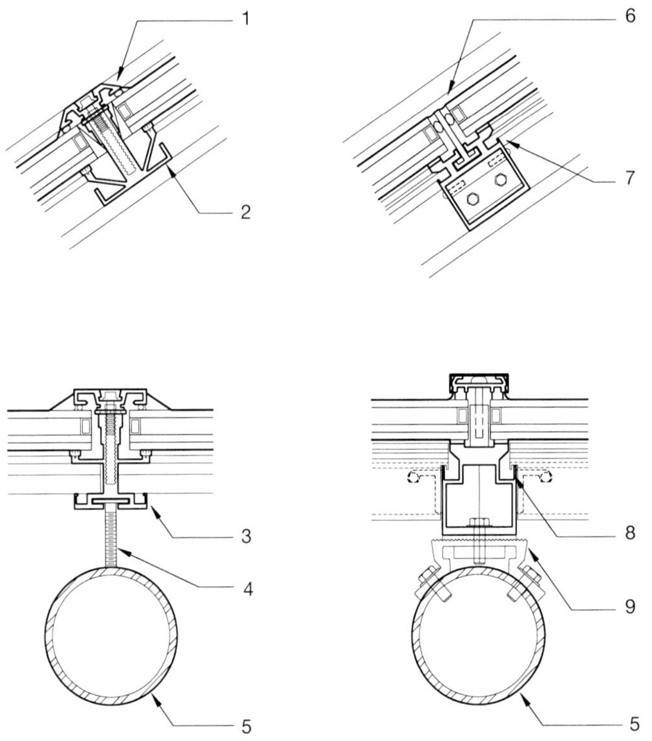

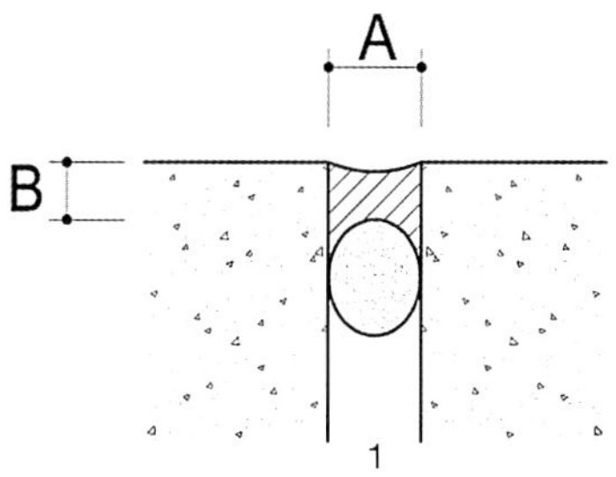

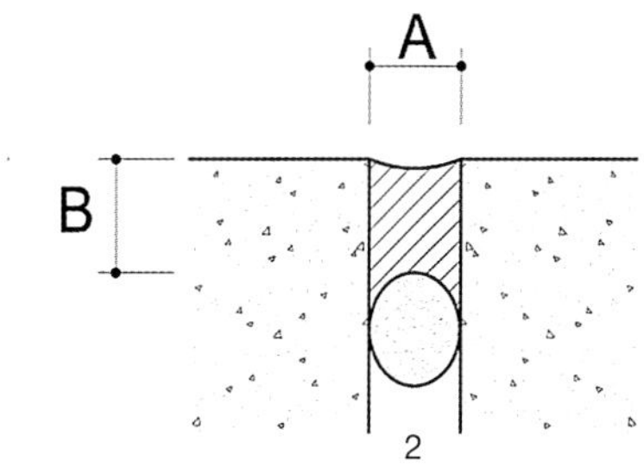

above Typical sealant detail:
1 Sealant width to depth A:B correctly installed, 2:1 minimum
2 Sealant width to depth A:B incorrectly installed, smaller than 2:1

above Gaskets prevent water from infiltrating the building. The enclosure of the 1995 Faculty of Law Building has a six-part nodal intersection of the glazing which is filled with an extruded gasket material. Faculty of Law, Cambridge University

center Sealant adhesion failure: sealant pulls away from substrate after joint movements have occurred.

right Triangular insulating glass units are mounted to the aluminum frame. Faculty of Law, Cambridge University

Joinery

Typical building enclosures consist of multiple panels of material joined together. The panel material can vary from stone to metal to glass and is rarely the source of water infiltration. Problems occur at the joints between panels. The control of such problems has remained the focus of the building industry from the advent of the initial curtainwall system.

Metal-to-metal and metal-to-glass joints should not be considered weather barriers. This transition will leak with time and should not be used as the sole deterrent against water infiltration. A chain is only as strong as its weakest link. If the sealant is not continuously installed through the complex window enclosure, air and water infiltration is inevitable.

Sealant failure can be cohesive or adhesive. Cohesive failure occurs when the movement capacity of the joints surpasses the designed movement capacity of the sealant. Cohesive failures can be avoided by using durable sealants and by detailing joints to accommodate anticipated movements. The joint depth-to-width ratio is critical. Typically, the width-to-depth ratio should be 2:1. Sealant manufacturers provide guidelines of how their product is to be installed and should be consulted regarding atypical conditions. Properly installed sealant joints should only have adhesion to two sides of the joint. Three-sided adhesion will prevent the joint from expanding properly. Backer rod or bond breaker material is used to ensure a joint can accommodate movements.
Adhesive failures occur when the bond between the sealant and substrates releases. This is displayed when sealant peels off metal, leaving no remnant of the sealant on the metal. Adhesion failures can be prevented on a building by testing the sealant's adhesive quality to anticipated substrates. It is not uncommon to use primers in order to allow sealant to bond to some substrates. Clean dry substrates are also extremely important to achieving good adhesion. The surface preparation of the substrate is critical to providing an affective seal. Primers are typically employed on substrates that require mechanical cleaning, such as brick, stone, and concrete. Most sealants should not be installed at temperatures below freezing or in damp weather.

Sealant substrate contamination and subsequent loss of sealant adhesion are difficult to avoid entirely, since invisible films of airborne debris can accumulate between cleaning, priming, and sealant application. Even wiping a surface with solvent to clean it prior to applying the sealant can lower its surface temperature, thus causing a film of moisture condensation that will impair or prevent sealant adhesion.

above Concrete enclosures can crack with time, providing a point of entry for water into the building.
Ontario Center, Chicago

below
1 Cracks in concrete enclosure are a point of entry for water.
2 Flashing at head of window provides a path for controlled water.
3 Baffled weep allows water to exit system.

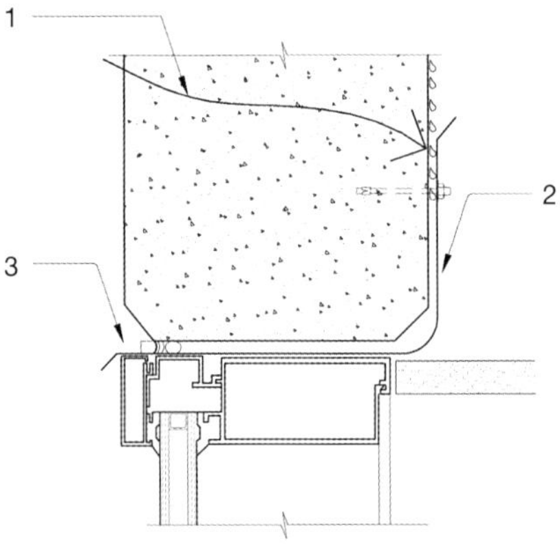

page 71 below The first pressure-equalized curtainwall enclosure in the United States: World Trade Center, New York

Dry Shrinkage

In addition to sealants, many of the outer seals of an enclosure are achieved through the use of gaskets. Dry shrinkage poses a problem to enclosure systems by creating a path for air and water leakage to occur. The gaps in the end of the extruded gasket can allow air and water to penetrate the system. Achieving better adhesion between the gasket and the metal can reduce the risk of dry shrinkage. Improvements in the formulation of the weather seal gaskets have aided this condition. For example, the use of silicone gaskets has reduced the effects of ultraviolet radiation on the joinery material. Silicone is not significantly affected by ultraviolet radiation and other factors of weathering.

Even a flawless seal installation is susceptible to problems. Most sealant and gaskets degrade with heat aging, exposure to water, strain-induced cracks, and ultraviolet rays of the sun. Even if a sealant joint is installed perfectly, adhering to both substrates, the ability of an enclosure to resist water infiltration will be limited. Problems such as hairline cracks in concrete, openings between metal-to-metal joints, and porous masonry substrates are unavoidable elements that allow water to bypass the sealant.

Sometimes the leakage is not due to the window system but to adjacent parts that allow for water to migrate into the building. Concrete and masonry surfaces are prone to crack and let water infiltrate the building with time. Sealant joints weather and allow for water to leak into a system. Condensation can accumulate on cold exterior surfaces of an enclosure. Properly designed systems will catch this water and direct it out of the building. Sloped concrete enclosures can be particularly difficult to protect because of concrete's inherent nature to shrink and crack with time. Even enclosure designs that effectively resist leakage generally include internal drainage systems that collect the water penetrating the outermost seals and redirect it to the building exterior. A secondary line of defense is the best protection against leakage into the building.

Second Line of Defense

Reliance on a single source of protection against water leakage is a plan prone to eventual failure. There are basically three methods used to prevent leakage to the inside of the building in a curtainwall or skylight system. The first method is to simply prevent water from penetrating the outer barrier of enclosure. The second method is to screen the water at the outer skin and provide an internal drainage system to carry any incidental moisture out of the building enclosure (cavity wall system or gutter and weep system). A third method of enclosure design employs pressure equalization. This type of enclosure has small holes in the outer skin of the building to prevent a pressure differential at the outer layer where extensive water is present.

Cavity Wall Construction: Water is expected to pass through the outer barrier. Penetrating water is directed through the enclosure system and flashed to the exterior at the base. This basic concept is illustrated in the conventional masonry cavity wall system. With time the bricks of a cavity wall will crack. Penetrating water runs down the inside surface of the wall and weeps out of the base flashing.

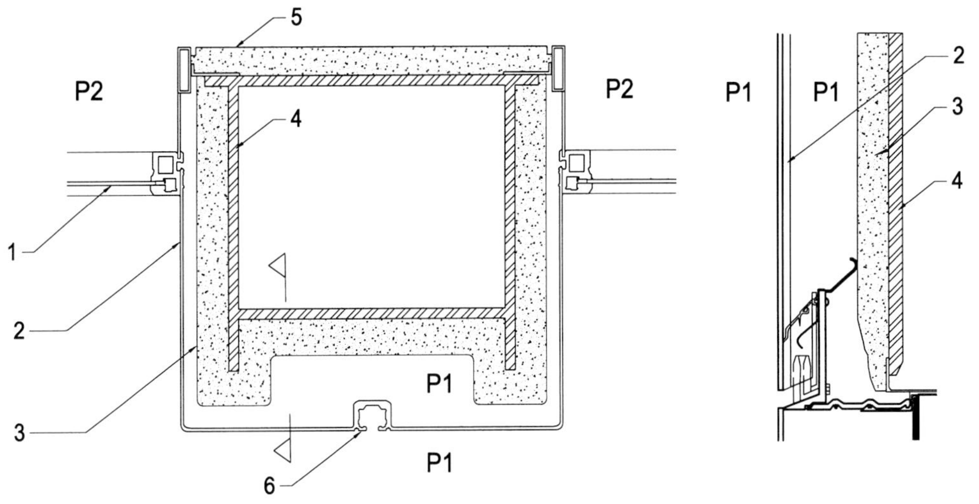

Plan and section of pressure-equalized enclosure
P1 Exterior air pressure (there is no change in pressure at the outer layer of the exterior skin)
P2 Interior air pressure
1 Window Glass
2 Aluminum Column Cover
3 Fire Proofing
4 Steel Column
5 Interior Wall
6 Window Washing Track
World Trade Center, New York

Gutter and Weep System in Skylights: The enclosure system most susceptible to water leakage are skylights. Unlike vertical walls, skylights are a cross between a roof and a glass curtainwall and must be designed accordingly. Because these systems traditionally rely on sealants for protection, they are prone to failure with time. Pressure equalization calling for the introduction of sufficient openings in the system to allow for the pressurization can be problematic in a wall that is close to horizontal. Similar to a cavity wall, today's skylight systems are designed with gutters and an internal drainage system to take penetrating water out at weeps. The gutters are used to collect water from condensation and leakage that typically forms on the underside of the glass and channel it to the exterior via the internal drainage system. Because the joints in a skylight have a greater chance to collect water than a curtainwall system, they are notorious leakers. Many people believe that all skylight systems will leak with time, and in a way they are right. The successful ones, however, have been detailed to take the leaking water out of the system.

Weeps and Baffles: Drainage holes are required to leak infiltrating water out of the system. Baffles are used to prevent water infiltration through the weeps by driving rain. Baffles can be made of a foam material but must be dense and secure to prevent insects from infiltrating the system and nesting. Spiders are notorious for spinning their webs across elegant façades on high-rise buildings. Pressurized cavities within the enclosure system provide insects with shelter from harsh natural elements like rain, sun, high winds, and snow. Baffles can prevent insects from entering weep holes. If the enclosure system cavity has loose construction debris or corroded metals, weep holes can be marked by stains. The design of these enclosure cavities must anticipate the entry of water and provide a clean method of draining.

Pressure-Equalized Cavity Wall Construction: This type of enclosure system attempts to reduce water penetration by eliminating the air pressure differential across the outer barrier. The outer barrier contains small openings to ensure that the internal cavity of the enclosure will have the same pressure as the exterior. Weepage at the base of these systems is critical to draining incidental water that can penetrate any of the small pressure equalization holes. This system must incorporate an effective air barrier on the inside surface of the wall to prevent water from being drawn into the building. Two key components to this design are an air barrier inboard of the outer surface and openings in the outer surface of sufficient size to allow pressurization of space between the air barrier and the outer layer of façade. This type of system prevents water from being sucked into a building by an internal pressure differential. Pressure-equalized design is particularly important in high-rise construction, where the enclosure system is susceptible to the forces of stack effect.

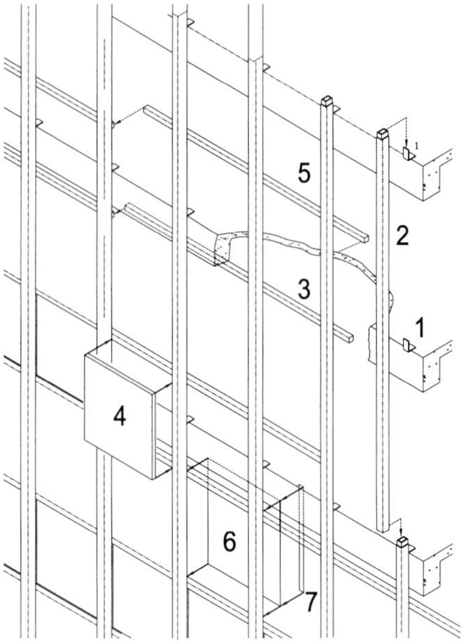

above Stick system:
1 Anchor
2 Mullion
3 Horizontal rail (gutter section at window head)
4 Spandrel panel
5 Horizontal rail (window sill section)
6 Vision glass
7 Interior mullion trim

right Unitized system:
1 Anchor
2 Pre-assembled framed unit

Stack effect occurs when high-rise buildings act like chimneys in the wintertime. Hot air will rise through these towers creating an inward draft through the lower floors. This flow of air creates an internal pressure on the enclosure system that draws air and possibly water through any holes in the exterior skin. Good seals at the perimeter will reduce the amount of air infiltration into the building. Providing air locks at roof hatches will also help to reduce the impact of stack effect.

Flashings are used to cover transitions, such as wall-to-wall, wall-to-roof, and wall-to-floor. Flashings are typically completed with a monolithic material like metal or rubber. Window systems' susceptibility to water infiltration can be greatly improved with a weeped through-wall flashing. This added detail will prevent infiltrating water from finding its way into the building. An upturned leg of the flashing will keep driving rain from entering the building.

Construction Methods

Rainscreen design, pressure equalization, gutter and weep systems, and cavity wall construction all have one common goal: prevent uncontrolled water from entering the building. There are many variations of these enclosure systems, and the way they are constructed can be as different as the means by which they shed water.

Stick Systems

Stick systems represent enclosures that are assembled in the field using individual components: mullions, gaskets, fasteners, and glazing. Stick systems can allow for considerable variation in situ. Because much of the material is cut to fit on site, this type of system can quickly be erected onto a building structure from elements shipped to the site. Stick systems are extremely dependent on quality control on the job site and are labor-intensive.

Unitized Systems

Unitized systems are factory-assembled units that are shipped to the site with all of the major components intact. Units are typically modular, one light wide and one floor tall, for attachment to the building structure with adjustable anchors. Unitized systems boast of better quality control because the units are assembled and glazed in the controlled environment of a factory. This is particularly important for four-sided structural glazing applications, where sealant adhesion is critical to holding the glass onto the building. Assembled units can be shipped across the country or around the world. Although unitized systems can take advantage of cheaper labor costs elsewhere, they are not typically less expensive than stick systems. There are many variables that affect the pricing of these two systems. The following are concepts that should be considered regardless of whether the enclosure is a stick or unitized system. Many systems are partially unitized, with the frames assembled in a factory and the glass installed at the job site.

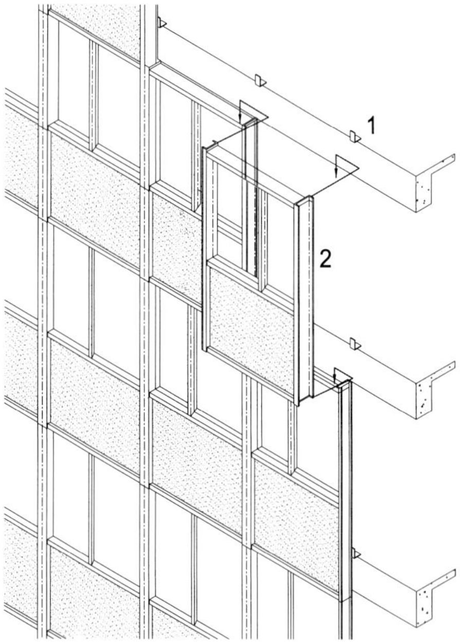

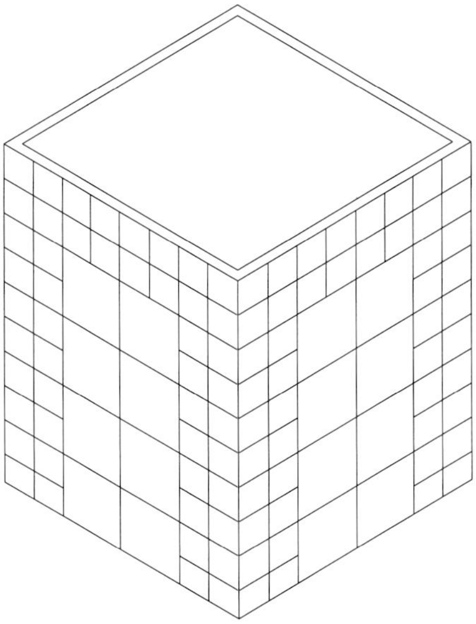

above Compartmentalization: The corners and the tops of a building are exposed to greater wind pressure and hence more rain. These areas should be compartmentalized at more frequent intervals to reduce the risk of overloading the internal weepage system.

right The mock-up testing was successful; however, the installation was not sufficiently compartmentalized, which led to leakage problems on the building.
Place St. Charles, New Orleans

Compartmentalization

A key component of either the stick or unitized systems is compartmentalization. The system's ability to weep water out of the building is dependent on limiting the enclosure area above the weeps. For example, weeps at the base of a 30-storey office building should not be expected to accommodate all 30 floors above. Compartmentalization is not just based on a building skin's surface area. With increased wind comes increased rain on a building. Because wind pressure is inconsistent over the entire building surface, building exteriors should be compartmentalized at more frequent intervals in high wind areas, for example, near the corners and the top.

Enclosure systems that cover large areas, for instance, high-rise buildings, require compartmentalization in order to ensure that internal drainage systems will not allow water to back up into the building. Place St. Charles, constructed in the mid-1980s in New Orleans, failed to compartmentalize the enclosure. The enclosure assembly was a pressure-equalized system that was successfully tested in a laboratory. Field installation did not separate the internal weepage system from floor to floor for the high-rise façade. This would have allowed for water to exit the system at various levels. The result was an enclosure that could be easily overloaded with small controlled water infiltrations over large areas.

Testing

One of the best methods of preventing leakage into the building is through testing of the enclosure system and its materials. Laboratory testing of the system typically includes tests for air leakage, water leakage, structural loads and thermal performance. Performance testing of enclosure systems was not always done. The laboratory testing of enclosure systems started in the 1950s and was a procedure developed for large enclosure assemblies of repetitive elements. The procedure greatly improves the probability of trouble-free enclosure performance. Mock-up testing is a learning process. Testing can uncover the need for minor modifications to the system, which can prevent long-term problems for the building. If not detected prior to fabrications, many defects can be difficult if not impossible to repair after installation.

What if the mock-up fails? It was money well spent! It is better to identify a problem at a testing facility than on the building. Modify the enclosure until it passes, document all changes, and execute all changes in the completed design. The mock-up is a tool and should be used as one. It may take several tries to pass the test; however, it is necessary to verify the performance of the enclosure.

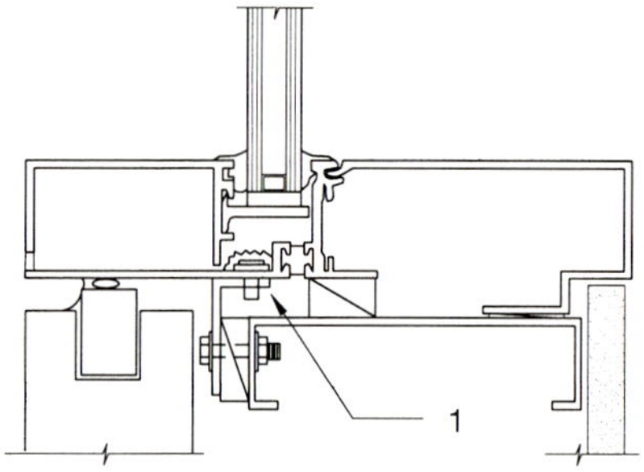

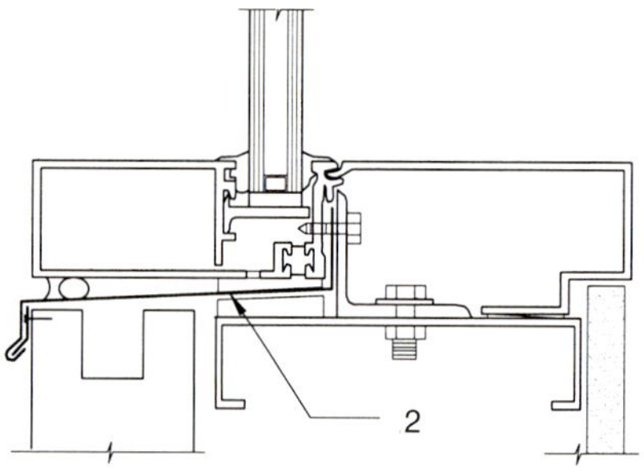

above Flashings
1 Curtainwall sill detail without back-up sill flashing
2 Improved curtainwall sill detail with back-up sill starter flashing

right Bug infestation: Insects can overwhelm building enclosures. Baffled weeps minimize the risk of insects nesting inside of the enclosure system.

Mock-Ups

Many think of mock-up testing as a method of verifying the enclosure's ability to meet minimum standards of performance. This is true, but it can also be the final design tool prior to fabrication of the system. Many modifications to the system can occur at the mock-up, which improve the performance and appearance of the enclosure. In the end, it is important that the tested assembly accurately imitates the actual design, and that it be assembled in the same manner as intended on the actual building. If possible, the individuals responsible for the erection of the enclosure on the building should be involved with the construction of the mock-up. While it is not possible to replicate every condition on the building, at least the size of the mock-up must be large enough to include all major elements and perimeter transition conditions. It is not usually necessary to repeat the actual building frame used to support the enclosure, but it is crucial that all actual details of the enclosure are incorporated into the mock-up.

Laboratory testing can be done to verify sound transmission, fire resistance, vapor transmission, resistance to seismic loads, overall condensation resistance, and thermal transmission; the major reasons for laboratory testing can be summarized as follows:

1 To verify the resistance of the enclosure against air leakage

The purpose of this test is to measure the amount of air leakage into the system. Industry standards identify a maximum of allowed infiltration per unit of area of an assembly. Excessive heat or cold into the building can tax a building's mechanical system. It is impossible for an assembly to be completely airtight. Air infiltration should be tested before the wall is wetted. Trapped water within a test specimen could reduce the detectable air leakage on a mock-up. Excessive air leakage in a building may cause physical discomfort and energy loss to the occupants.

2 To verify the effectiveness in controlling water penetration

Resistance to air and water infiltration can only be determined through testing. The test intends to simulate an enclosure subjected to high winds combined with heavy rainfall. The assembly is subjected to a predetermined internal pressure to simulate the force of wind on the enclosure. Water is sprayed on the assembly at a specified rate for a given time period. Failure of the test occurs when water appears on the inside surface of the enclosure during the course of the test. Controlled water inside the system that weeps to the exterior is allowed. A second method of testing for water penetration into a building is the dynamic pressure test. Dynamic testing is used to determine if water will infiltrate an assembly if exposed to a pulsating pressure. The in and out turbulent flow of air as occurs during a storm is typically simulated by using an air plane propeller engine in conjunction with water spray racks. Because of the turbulent nature of the pressure, water may penetrate a system, which would not show up in the uniform static air pressure test. It is recommended that the water infiltration test should be conducted after structural loading to confirm that minor deflections in the system will not trigger leakage into the building.

right Laboratory mock-up testing: Dynamic water testing uses an airplane propeller to simulate the pulsing action of a driving windstorm.

3 To verify the structural adequacy of the enclosure under wind loads

The strength and stiffness of framing members may be determined by engineering analysis and calculations; however, the structural performance test is the final check to find out if the composite assembly meets the design loads. Pressurizing a test chamber encased by the enclosure system being tested simulates wind loads. Inward and outward forces are achieved by drawing air in or out of the chamber. Failure of this test consists of fracture of any individual element or deflection beyond the established maximum under loads equal to or less than the design loads.

If the proposed building is in a climate that has severe temperature differences, it is recommended that the above testing be repeated after exposing the assembly to a representative temperature range. Thermal cycling can trigger defects through expansion and contraction of the system. It is also a good idea to test the system for air and water infiltration after the testing for structural performance. The deflection of the enclosure under structural load can simulate the real movement of the building and uncover a weakness in the system. Overload testing to demonstrate the safety factor of an enclosure design is usually the last test conducted on the mock-up. These tests help designers know how the system performs in unrealistic circumstances and provide additional comfort in knowing that the system can resist loads beyond what is was designed for.

above Laboratory mock-up: Construction Research Laboratory in Miami, Florida, has been testing building enclosure since the 1950s.

right Laboratory mock-up testing: Water infiltration testing includes a hose rack to evenly distribute water over the entire mock-up surface.

Laboratory testing is typically conducted at an independent testing facility in order to ensure impartiality in administering and evaluating the test. Testing of enclosure systems is relatively new in the construction industry. The first recognized facility in the world was the Construction Research Laboratory in Miami, Florida. Today, this facility is the world's largest testing ground for building exteriors. A.A. "Sak" Sakhnovsky founded the company in 1968. From New York's World Trade Center to the Petronas Towers in Kuala Lumpur, Construction Research has given the industry the ability to analyze enclosure systems prior to building them. Mock-ups, which can be up to 70 foot (24.3 m) tall, are subjected to winds as high as 311 mph (500 kmh), rainfall equal to eight inches (20 cm) an hour and temperature swings from -20° F below zero to sweltering 180° F (-29° C to 82° C).
A.A. Sakhnovsky has identified water infiltration as the number one factor in failed tests. He has noted that 90 % of curtainwalls fail the first water test. Modifications are made until the mock-up passes. Testing is a cost-effective method of verifying the design of new systems prior to assembly on site.

above Laboratory mock-up testing: Failure of 1/2 inch (12.7 mm) tempered glass at 90 psf (4.3 kPa) pressure during negative structural testing.

Field Installation

Laboratory testing can help identify major design errors and various installation challenges, such as sequencing, building tolerances, and workmanship. It is advantageous for the workers involved with assembling the laboratory mock-up to be represented later on the actual site to supervise the installation of the enclosure. For custom systems, this may be a requirement for the project. Mock-up testing cannot verify many quality control measures, which are carried out on the job site – for example, when installers are pushed to complete work on the building quickly in adverse weather. Nor can enclosure designers confirm what happens to the material on the way to the job site, for example, the racking of frames in transport or handling. For all of these situations, it is recommended to provide random testing in the field to ensure the system on the building is as good as the system that passed the laboratory tests. Field testing is not exclusive to large-scale projects and can be as simple as spraying the exterior joinery at random locations with water from a hose and looking for seepage into the building. Points of leakage can thus be identified and corrected.

Field installation is crucial to the success of an enclosure. Pressurized air and water infiltration testing can be achieved in the field by constructing temporary chambers on the inside surface of the assembly. Air is drawn through the system while water is sprayed simultaneously onto the outside of the enclosure. One of the simplest and yet most effec-

right Field hose testing: Field testing is a method of checking the quality of the installation, the materials used, and the design of atypical transitions.

page 78 Field installation is critical to the performance of any enclosure system.

tive methods of testing in the field is a standard hose test. This is a water spray test conducted with a garden hose and special nozzle. It permits detection of leakage due to faulty fabrication or installation. Both tests are monitored from the inside of the building to determine if water leaks through. If leakage occurs, the system must be modified and retested. Water testing should not be conducted during freezing temperatures or while it is raining.

The most critical time to complete field testing is at the beginning of a project. Early testing can eliminate these faults before the remainder of the installation is completed. Field testing when the first units of an enclosure are installed can uncover any faults with the assembly of the system. Random field testing of the system as the project progresses will help verify that quality control has been maintained throughout the progress. Atypical conditions not part of laboratory testing can be reviewed for air and water infiltration. Field

above Innovative glass roof enclosures need sophisticated glazing systems. With these unique systems, water testing is a necessity.
Milwaukee Art Museum

below View of skylight from below.
Milwaukee Art Museum

testing is also important to ensure that transitions to adjacent systems have been properly executed. Random sealant adhesion field testing is important to verify that substrates are being prepared properly. These tests confirm that the seals are being installed as designed and that the sealant material is curing properly.

How can Leakage be Avoided?

Design drawings for enclosure systems typically stay in the two-dimensional world of plan and drawing sections. However, since water moves laterally within enclosures, finding the path of least resistance, these drawings can limit the review of water infiltration weakness. Isometric drawings of corner and transition joints help tell the story of how seal continuity of an enclosure is obtained. No matter how impressive a building enclosure system may appear on paper, a laboratory test of the typical details is a useful tool in uncovering unwanted problems that could be found in the field. Laboratory testing includes a review of the thermal, structural, air, and water performance of an enclosure. It also helps to confirm erection tolerances and sequence of installations. The most common type of failure for mock-up testing is water infiltration; however, designers should not be discouraged by this fact. Laboratory testing should be seen as a tool in the design process. Modifications in the design are far easier to achieve in this setting than after the enclosure has been installed on the building.

Joinery is typically the source of air and water infiltration. Proper preparation of substrates and installation of sealants is critical to maintaining a weather-tight seal. Perfect seals are difficult to achieve even in a factory setting.

Time and weather will adversely affect sealants and gaskets that make up joinery details. The best plan for keeping water out of the building is to provide a second line of defense. Back-up systems like through-wall flashings are useful in draining infiltrated water out of the system. Techniques like pressure equalization bring primary seals away from the outside surface, protect from the environment, and reduce the amount of air and water on these joints. Field testing of random areas of the enclosure will ensure that the materials and installation techniques are consistent with the laboratory test. Field tests can also be used to verify atypical transition details that were not tested in the design process. Although not always apparent to building owners, advances in enclosure system designs have improved the performance of our building skins. As our buildings become larger and more technically advanced, new materials and techniques will protect the building inhabitants from leakage.

Lessons Learned

1
In order to minimize the risk of water infiltration, provide a secondary line of defense. The exterior seals of building enclosures are going to leak. The question is: where does the water go once it gets past the first barrier? Properly designed and installed systems direct the water back to the exterior.

2
Do not rely on a single line of sealant to prevent leakage.

3
Large areas must be compartmentalized to prevent an overload on the system. Areas that receive greater amounts of wind are also susceptible to greater amounts of water and will require additional compartmentalization.

4
Concrete façades should have properly designed and constructed control joints in order to minimize shrinkage cracking.

5
Joinery is the most common source of leakage. All exterior sealant joints must be detailed to accommodate movement and adhere to adjacent substrates. Sealants should be tested for adhesion and compatibility prior to installation.

6
Although mock-ups and material testing are important to a design, they will not change the fact that sealants are vulnerable to workmanship installation in the field and deterioration occurring years after construction has been completed. The best systems have seals that are installed in the shop and not exposed to weather in the final system. This type of joinery can be found in unitized pressure-equalized enclosures. Unitized systems reduce the risk of faulty installation in the field because they are completed almost entirely in a factory setting. Pressure-equalized enclosures provide a primary seal inboard of the outside face of the building. These seals are protected from the environment by a cavity that maintains the same pressure as on the outside of the building.

7
Laboratory and field tests are the best methods of ensuring a leak-free enclosure on a building. Ideally, for custom systems, the same workers that construct the project should construct the mock-up.

8
The successful prevention of leakage in enclosures is dependent on the workmanship of the installation.

Energy

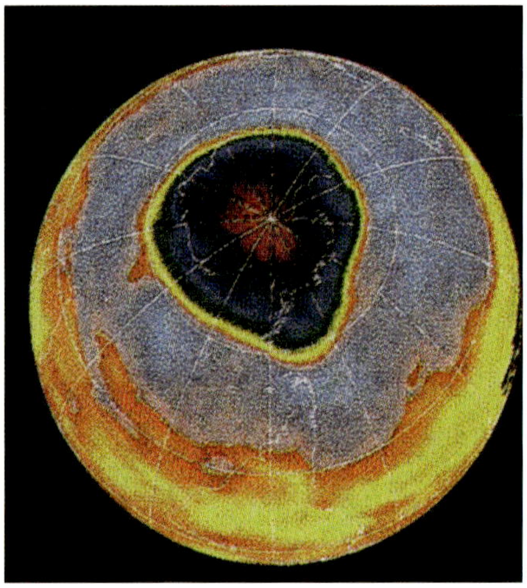

Buildings, like our bodies, exchange moisture, heat, and air with the exterior environment. The building envelope, like our skin, is a buffer zone critical to protecting the contents within. Thermal barriers prevent the transfer of heat, vapor barriers prevent the transfer of moisture, and shading devices limit the sun's glare and harmful ultraviolet rays. Building skins with poor barriers allow for these undesirable elements to adversely affect our interior spaces. Because so much of a building envelope is made of glass, the performance of this material is critical to the comfort of the building's inhabitants. When a building is uncomfortable at its perimeter, the problem is typically related to the glazing enclosure. When condensation forms on the exterior glass, obscuring views and damaging interior surfaces, the building envelope is criticized. When direct sunlight limits the use of interior spaces because of solar radiation and glare, the windows are to blame. Many innovative cladding systems have been developed to combat these natural forces. After their successful installation, questions have been raised as to their effectiveness.

In addition to preventing condensation, glare, and the transfer of heat, a successful building enclosure will reduce the amount of energy required to keep a building's inhabitants comfortable. Buildings have a significant impact on the world's resources and on the health of people who inhabit them. Approximately 45 % of the carbon dioxide deposited into the atmosphere is due to buildings. In a responsible effort to reduce these pollutants, building skins have evolved into sophisticated systems that reduce the amount of energy required to sustain a building. In some cases, these complex enclosures resemble active machinery. Some have moving parts controlled by computers to transform the glazing in response to changing exterior climatic conditions. The problem with many of these intelligent glass façades is their added maintenance and complexity. In many cases, these shortcomings have challenged the success of the technology. Improving our building skin's performance

above Roughly one half of the earth's ozone damage is caused by chlorofluorocarbons (CFC) used in air-conditioned buildings.

right Double-glazed façades reduce the need for cooling our buildings.
DGAC Building, Paris

page 83 Double skin façade
DGAC Building, Paris

is a critical task in providing the best energy efficient solution. New technologies have stimulated researchers into rethinking how we can cover our buildings while allowing the much-desired light to come in. Energy efficiency in building envelopes has evolved from insulating glass units to double skin glass façades. Active and passive shading devices are employed in order to improve the comfort of the inhabitants within. In the process of developing these new methods of energy conservation, many enclosures have had difficulties. Occasionally, the quest for energy savings has led building owners to envelopes that have failed.

Completed in 1995, Galeries Lafayette, Berlin, was designed as a mixed-use, retail and office building that has a sophisticated skin designed to conserve energy. The 30 meter high elevation has a double skin façade curtainwall with backlit spandrels, designed as advertising boards. In addition to its function as a weather barrier, the double skin façade allows for natural ventilation to the offices behind them. The building top is shaped as a series of truncated cones made of curved fritted glass panels inclined backwards. The retail area is on the ground floor level with shops, offices, restaurants, and apartments on the upper levels. The double skin façade consists of an outer 12 mm (1/2 inch) sheet of toughened glass supported by point-supported lights. The 30 mm (1 1/5 inch) interior insulating glass units has an 8 mm (1/3 inch) outer pane and a 6 mm (1/4 inch) inner pane with a low-E coating and argon gas filling. Perforated louver blinds of stainless steel are integrated into the 20 cm (7 3/4 inch) wide cavity to improve the thermal performance of the wall. Air can circulate in this space through 15 cm (6 inch) slits above and below the outer light of glass. 2.75 m high and 1.35 m wide (8 x 4 foot) glass as fixed glazing and turn-and-tilt windows allow for natural ventilation. The outer light of glass has a silkscreen pattern on the inside surface at many locations. The 20 % frit of gray dots increases to 100 % frit in some areas of the façade. The interior insulating units are not curved and provide a segmented circular profile in plan. Many of the outer lights of glass are curved in order to follow the plan of the building. Much of the exterior glass underwent two-heated processes prior to installation: one for the curving and one for the ceramic frit application.

page 84 above A sophisticated ventilation system incorporates the building façade with the mechanical system. The target is to reduce energy consumption by about a third of what is typically required by an office building.
The New Parliament Building, London

page 84 below The exterior lights of glass were removed and replaced after a number of them failed.
Galeries Lafayette, Berlin

right This double skin façade was reclad in 1998 due to a problem with the exterior light of glass.
Galeries Lafayette, Berlin

below Detail of double skin façade:
1 The exterior glass light of the entire double skin façade was replaced due to breakage.
Galeries Lafayette, Berlin

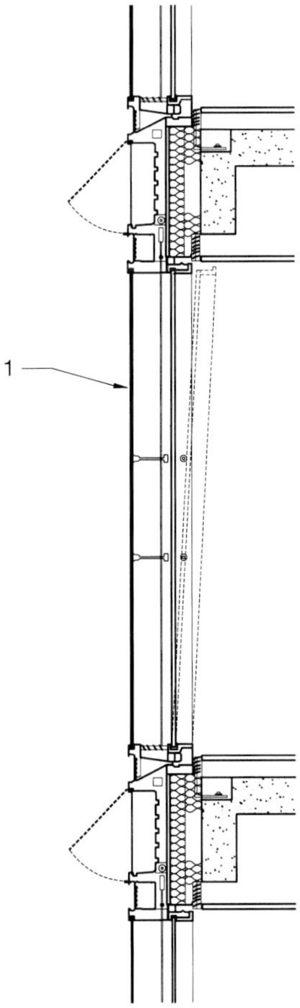

The 8,000 square meter envelope suffered a serious setback when the individual lights of the double-glazed façade started to displace in 1998. Although the occurrences were limited, it put the entire exterior glazing in suspicion. At enormous expense and energy, all of the outer lights were removed and replaced. Modifications to the replacement system included the use of laminated glass for the outer skin. The double skin façade failure of Galeries Lafayette represents a recent entry into the evolution of energy efficient enclosure systems. In order to understand how these systems work and what their weaknesses are, one must review the history and evolution of the double skin façade.

Double Skin Façades

As the industrial world became aware of the energy crunch following the oil crisis of the 1970s, there was a brief reduction in the use of glass to enclose our buildings. The large expanses of glass on many of the buildings at the time were criticized as causing tremendous heat loss in winter and overheated interiors in the summer. These buildings required large energy consuming air-conditioning systems for heating and cooling. As a consequence, the glazing industry developed colored and reflective coated glass to obstruct the sun's rays. This, unfortunately, led to darker interior spaces requiring additional electric power for interior lighting. The reduction of natural light was a sacrifice with which many building occupants were not happy. Daylight in the work environment was a positive benefit to the people inside the building. The architectural community was faced with a new challenge, and one of the answers was the development of the double skin façade.

The Hooker Chemical Building constructed in Niagara Falls in the early 1980s was one of the first modern prototypes of the double skin façade building enclosure. The design ended the search for a transparent façade that was energy efficient. Two walls of glass were used to separate the interior offices from the exterior environment. Since the construction of the Hooker Chemical Building, double skin façades have become very popular throughout Europe. From the DGAC (Direction Générale de l'Aviation Civile) Building in Paris to the Triangel Office Building (1996) and GSW Extension (1999), both in Berlin, double skin façades are being constructed on south-facing walls as a responsible method for reducing building energy consumption. In addition to providing a range in architectural design possibilities, a double skin façade has the following benefits:

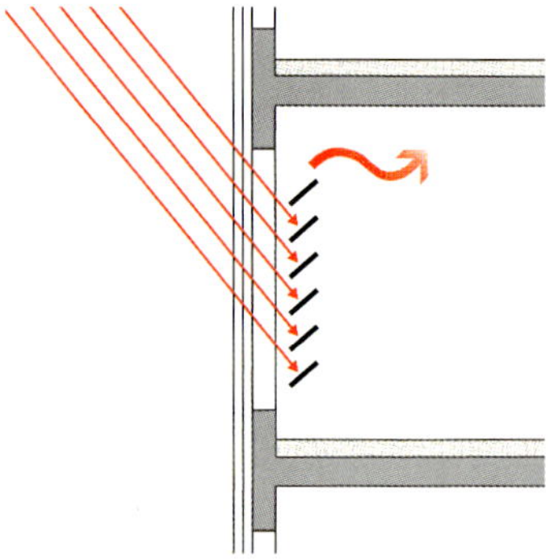

Traditional glass skin

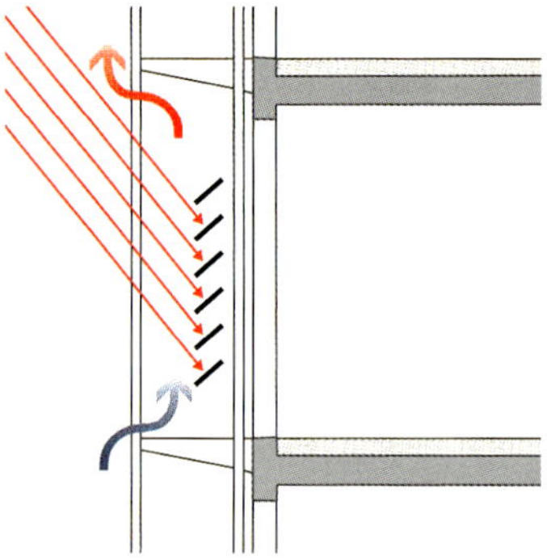

Double skin façade
If desired, the interior glazing stays cool

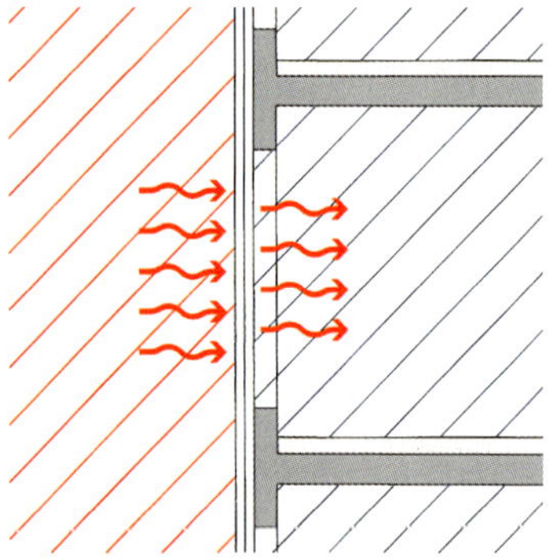

Traditional glass skin

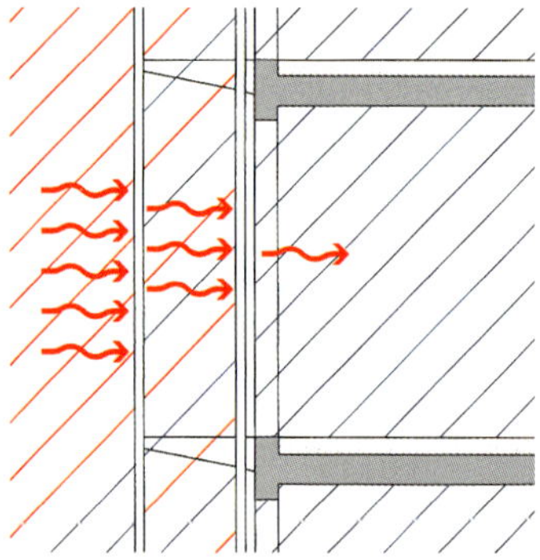

Double skin façade
Improved thermal performance

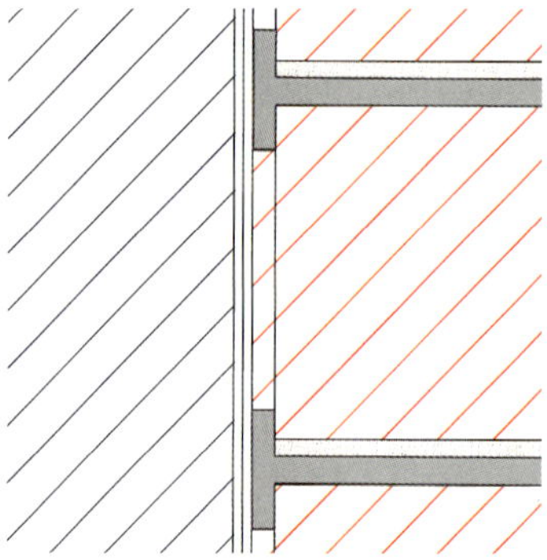

Traditional glass skin

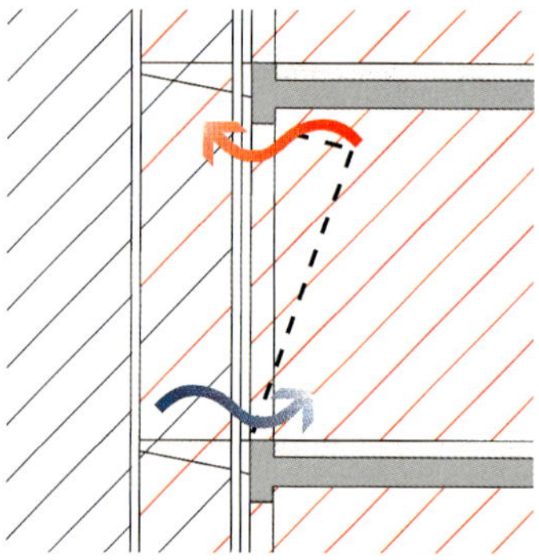

Double skin façade
Controlled natural ventilation

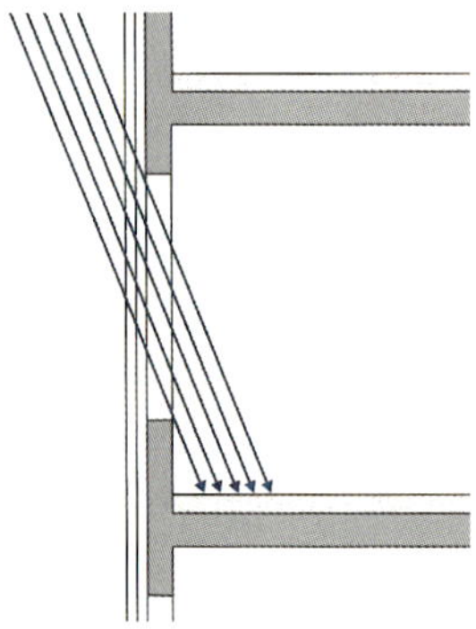

Traditional glass skin

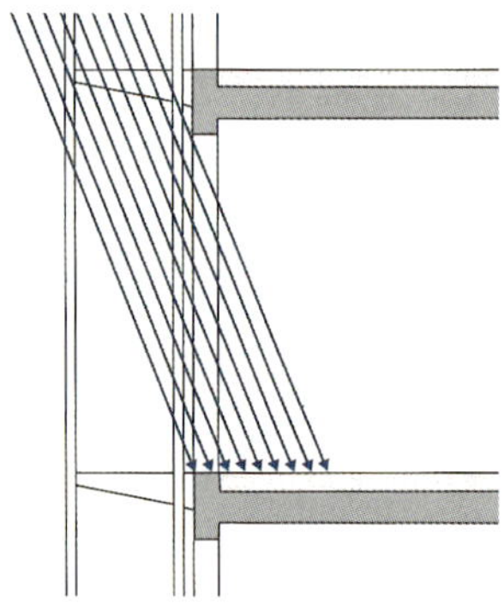

Double skin façade
More natural sunlight

above Double skin façade
GSW Extension, Berlin

right Double skin façade
Triangel Office Building, Berlin

Radiation: Double skin walls are intended to improve the comfort level near the enclosure. Double skin façades decrease the transmission of heat. Most people feel comfortable at a room temperature of 72° F (22° C). However, if they are surrounded by large surfaces that are at lower or higher temperatures (such as a conventional double-glazing) they experience discomfort due to the radiation effect from the adjacent surfaces. Double skin façades greatly reduce the effects of thermal radiation at the perimeter of a building. The occupants can locate workstations adjacent to the glass with minimal discomfort.

Natural Ventilation: Natural ventilation for a high-rise building can be problematic because of high winds. Double skin façades make ventilation possible by screening high winds with the outer layer of glass. Air can naturally ventilate through the two skins and provide fresh air to the inhabitants by opening the inner light of glass. Depending on the envelope's design parameters, a double skin façade has various methods for controlled ventilation.

Noise Reduction: Double skin façades dramatically improve sound insulation. Users can actually receive the benefits of open windows while still maintaining some level of acoustical separation.

Energy Conservation: Double skin walls are designed to improve a building envelope's performance by reducing cooling load and air change rates. The low U-value of double skin façades allows for larger glass areas compared to conventional double-glazing.

Improved Lighting: Because of their superior thermal performance, double skin façades allow for more glass and, therefore, more daylight in the rooms. This increases the light in the interior and enhances visual comfort while reducing the consumption of artificial lighting energy. This is particularly advantageous on overcast days.

Improved Control: A controllable double skin façade can stop negative external climatic effects from influencing the inside of the building while still allowing for many positive external climate effects. Double skin faces allow for nighttime cooling of the building without sacrificing security.

In summary, double skin façade designs are intended to improve the quality of the workspace at the perimeter of a building. In addition, double skin façades have solved many of the energy concerns of glass enclosure systems. Double skin façades are an innovative advancement in enclosure design; however, they are susceptible to the same types of glass failures as conventional glazing. In fact, double skin façades are more prone to failures because they inherently have twice the amount of glazing as a conventional façade enclosure.

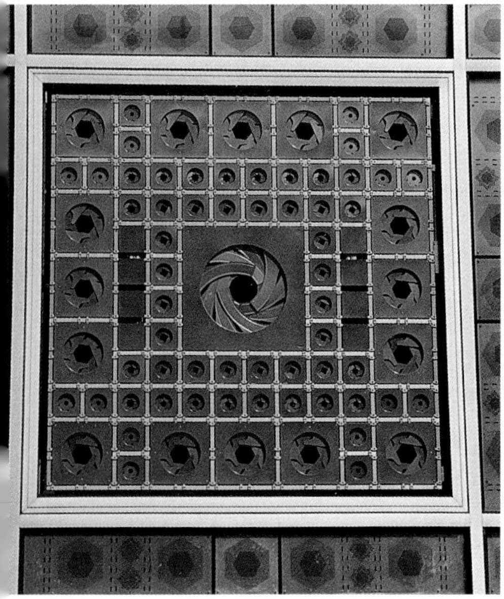

above Shutters on the façade control the light levels within the building.
Institut du Monde Arabe, Paris

below Many of the shutters no longer work and are fixed in the open position. The darker window indicates that the outer insulating glass unit has a failed hermetic seal.
Institut du Monde Arabe, Paris

page 89 Interior spaces include a public library and art gallery. Both of these spaces require very controlled light levels.
Institut du Monde Arabe, Paris

Intelligent Glass Façades

Parallel to the development of the double skin façade was the concept of dynamic building enclosures. Developed through the 1980s, these systems were designed to respond to occupants' needs to provide protection from the sun and additional lighting automatically. In practice, the concept of intelligent glass façades is often coupled with complex technology. Not only are our building skins expected to stop rain and wind, computers and sensors are controlling their components in order to respond to changing environmental conditions.

Completed in 1987, the south façade of the Institut du Monde Arabe in Paris became the prototype for façades that would change with their environment in order to maintain the desired interior environment. The shutters on the façade were intended to reduce solar gain along the south side of the building. There is a mix of functions behind this façade, including a library, a museum gallery, conference rooms, and private offices. Thousands of iris-like mechanisms create a unique texture across the façade. The windows resemble the carved pattern found on some ancient Arab lattice work (meshrabiyehs). 27,000 aluminum shutter elements open and close by means of an electro-pneumatic mechanism, regulating daylight transmission between 10 % and 30 %.The modular units were designed with photoelectric cells that control the shutter mechanism. Built according to a principle similar to the double skin façade, the 1.8 x 1.8 m (5 x 5 foot) panels with light-sensitive diaphragms are set between an exterior fixed insulated glass unit and an operable monolithic light of safety glass that opens like a door for maintenance.

Plagued by a strained maintenance budget, many of the modular elements have gone unrepaired. The shutters behind the insulating glass units no longer work as they did when the building opened in 1987. Rather than operating by photoelectric cells, the shutters are now controlled by a computer. Many of the devices are inoperable and are left in the open position, no longer changing with the exterior environment. These windows were the forerunner to many of the shading devices used to control light levels of later constructed curtainwalls. Many of the problems with these types of sophisticated enclosure systems were not left on the banks of the River Seine. Shading devices have continued to aggravate architectural façades in projects after the Institut du Monde Arabe.

right Exterior shading device
Fondation Cartier, Paris

below Shading device controls do not work above the main entrance.
Helicon Building, London

page 91 right Fixed glass louver blades
Heathrow Airport, London

page 91 left Sophisticated double skin façade system:
1 The shading device controls do not always work.
Helicon Building, London

The Helicon Building, completed in 1998, in London was a solution to many of the problems of an energy efficient office in an urban site. City areas have poor air quality which undermines any natural cross ventilation. The Helicon Building is an urban solution with automated/ventilated double skin with electrically controlled shading devices for solar control. The façade is highly glazed to allow for maximum use of daylight. The solar screens that double as light shelves prevent excessive heat build-up and deflect daylight into the center of the building. Aided by the specialized façade system, the building combines mixed-mode ventilation and air conditioning.

The Helicon Building is a wonderfully detailed building that aggressively tries to combat environmental forces with active technology; however, the shading devices and double skin façade vents do not always operate as designed. The shading devices are controlled by sensors, which can be easily damaged. Many of the sensors which control the devices have been obstructed or inadvertently tampered with after the completion of the project. Override switches also appear to be utilized by the occupants, preventing the shading devices from performing as designed. It is possible that complex enclosure systems which are turned over to people with limited understanding of how the system works can jeopardize the success of that system. All of these problems undermine the logic of installing fragile, complex, high maintenance elements on the exterior of a building.

Sun-Shading Devices

The traditional concept of interior shades and drapes has recently been challenged because these devices do not protect the outer glass surface of an enclosure from the sun. A new philosophy in enclosure design locates the shades on the outside of buildings. This technique attempts to keep the glass and the building skin cool when exposed to direct sun light. Fabric shading devices have been installed on the exterior and are controlled from the interior. A fundamental problem with this solution is that the shades are susceptible to wind, rain, and snow, and require greater maintenance. The shades can be designed to rise automatically when sensors detect high winds. This added level of complexity plus limited use during windy days seriously challenged the shading concept's effectiveness. Fixed metal and fritted glass sunshades are used in façades with high sun exposure and are less susceptible to damage during high winds. These elements are changing the appearance of our buildings. The north façade no longer looks like the south as protruding sunshades on the outside of a building typically appear only on the façades that face the sun's path. Although these fixed passive devices appear more durable than fabric sunshades, they are still susceptible to damage and do not change with the variation in light levels on the exterior.

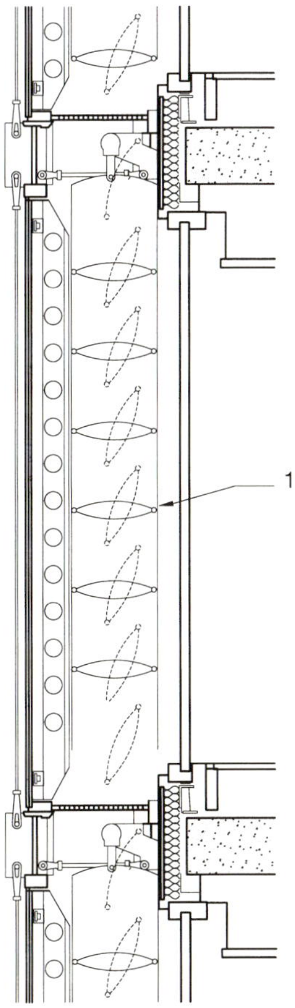

Glass Sun Shades

Rotating glass sunshades have the ability to change with exterior conditions and interior needs. This system was developed to reduce the internal room temperature of the building and increase daylight in perimeter spaces. The glass sunshades have the added benefit of reducing the amount of electricity required to light interior spaces. In times of insufficient daylight such as with an overcast sky, the louvers can be moved into position to reflect daylight into the building. Low-E coatings (scratch-resistant) or partial frit patterns are used to increase reflectance. While maintaining views to the outside, the fully closed louver blades reduce the cooling loads produced by solar energy. Unlike other external shading devices, the louvers remain fully operational during strong winds and they can be altered depending on the building inhabitants' needs. Automatic controls run on low energy to control the shade's orientation as the environment changes. The controls are part of a building management system (BMS) that supports the building's heating and cooling needs.

This building in Bordeaux has a damaged fixed metal shading device.

Sunshades are not only used with glass but also with acrylics. Completed in 1992, the Siemens Pavilion used automated full-height solar shields made of prismatic acrylic panels to deflect the sun rays. Built as a Pavilion for the Expo '92 in Seville, the façade is designed with full-height, segmented solar shields suspended from the roof construction. The 17 meter high and 28 meter wide (49 x 80 foot) shield moves around the circular pavilion tracking the position of the sun. Mounted on the shield are moveable horizontal louvers made up of 25 x 250 cm (10 inch x 7 foot) prismatic acrylic panels. These panels reflect direct daylight while admitting diffused light. A central electronic control system governs the tracking of the shield around the building. The energy required to operate this system comes from solar cells mounted on the roof. The ever-changing position of the shield and the angle of the louvers continuously give the pavilion a new appearance.

Like other intelligent façade systems, these automated glass sunshades have serious maintenance issues with controls, motors, and pivots. Although the devices have an added benefit of providing additional lighting at certain times of the day, it is difficult to automate and predict the needs of humans. During holidays and weekends, the devices continue to operate in full force, even though the building may be completely empty. The shades concept struggles with the goal of using energy as a means of saving energy. Although the diffuse lighting created by the glass louvers may reduce some level of internal lighting, it will never provide the amount of intense light required for most tasks. This is an interesting yet expensive architectural element used to bring natural light into buildings and decorate exterior façades. While maintaining a level of transparency, the glass louver system cannot always be used as a substitute for internal shades. Additional sunshade devices like blinds or drapes may still be required to reduce the glare of the low sun or to completely block out the sun if desired. Projects have attempted to reduce the need for internal blinds by providing a full frit on the glass. Unfortunately, this presents a visual obstruction to exterior views. Like all moving glass ele-

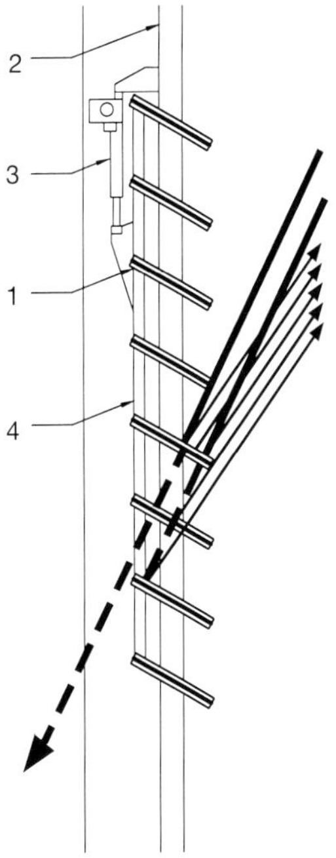

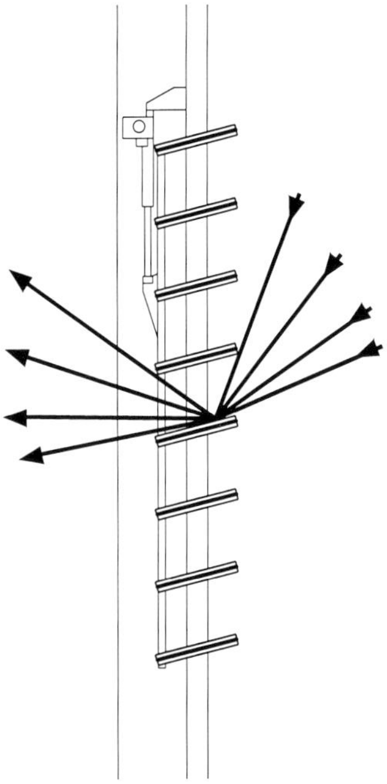

above left Glass sunshade detail
Hypo-Bank Ost, Munich

above right Glass louvers reflect the sun's rays into or away from the building:
1 Frameless pivoting louvers.
2 Louvers are connected to. mullion with stainless steel connections.
3 Louvers are powered by a linear motor.
4 Louvers are connected with a driving rod to allow for movement in sequence.

center right Broken laminated glass louver blades
Cité International, Lyon

center right Motorized glass louver blades influence light, air, and temperature quality of the interior.
Cité International, Lyon

ments, the fins are susceptible to breakage and are often laminated to prevent fallout. The production of laminated glass requires an intense amount of energy and expense. The rotating glass sunshades are a response by the architectural community to the growing need both to reduce the amount of energy our buildings need and to increase natural light into the building. They represent just one developing tool in a larger movement toward sustainable building design.

Sustainable Building Design

The concepts of sustainable building design concentrate on minimizing the amount of energy used to operate a building. The typical office in the UK consumes 200 kWh/m^2 (19 kWh/ft^2) over a typical year. Most sustainable building design projects aim to reduce this to 100 kWh/m^2 (9.3 kWh/ft^2). Some buildings have achieved energy levels of 75 kWh/m^2 (7 kWh/ft^2) through the use of photovoltaics and higher levels of construction. Many projects in Europe are trying to achieve these levels with the use of natural ventilation, nighttime cooling of building structures, and more low-technology building methods. This type of design philosophy is seen as a critical step toward preserving our environment. The movement has many innovative concepts that affect the design of the glass enclosure. One of the basic principles developed is the use of interactive window systems (windows designed to open at night to cool the concrete structure) linked to BMS (building management systems).

The EDF Eléctricité de France office building in Bordeaux, completed in 1997, was designed with exterior windows on a BMS system and exposed concrete structure for nighttime cooling. The top windows of each floor were designed to automatically open at night during the summer months to cool the exposed concrete structure. When the system was initially used as designed, it was found that opening the windows created some problems. In addition to being a breach in security, opening the windows at night brought too much humidity

above and right Operable windows designed to be controlled by building management system to allow for nighttime cooling.
EDF, Bordeaux

page 95 Fixed louver blades
New Parliament Building, London

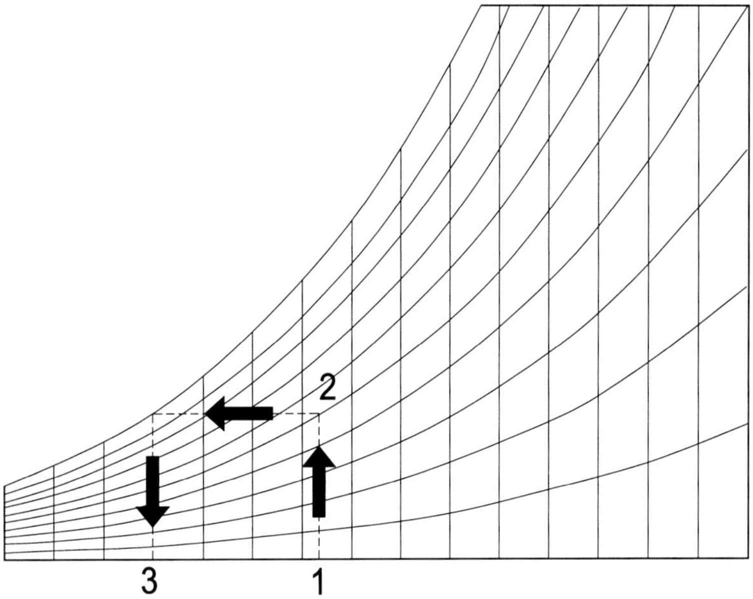

above Simplified psychometric chart for determing the dew point of an enclosure:
1 Temperature of the room
2 Humidity level of the room
3 Temperature at which condensation will occur (dew point) in the room

above left Condensation at an indoor pool can lead to freezing of emergency exit doors

right Mechanical floor ventilation of monolithic exterior glazing at the lobby
Sony Center, Berlin

page 97 Bibliothèque Nationale, Paris

into the building and caused the large copy machines to malfunction. The building has many energy saving features that make it a successful sustainable design; however, the end result of the façade is that the building is constructed with an elaborate window control system that is not fully utilized by the occupants. Humidity is a factor that must be considered in designing all building enclosure systems.

Condensation

Because of its low thermal resistance, glass is susceptible to condensation and frost. We are all very familiar with this phenomenon. Moist summer air in contact with the cool surface of a glass container holding an ice filled drink will condensate. Certain building types are prone to the formation of condensation. Indoor pools in colder climates are an extreme example. In hot humid conditions, cool inside surfaces are created with the use of air-conditioning. As hot humid air comes in contact with a cool glass window surface, condensation can occur. This phenomenon can also occur in the reverse condition. A cold exterior climate can produce a cold exterior enclosure. If the interior space has a high relative humidity, the moist vapor in the air will condense on the glass. A clear example of this can be seen in enclosure systems of monolithic glass around a swimming pool in the winter. Once the moist air of the pool comes in contact with the cool glass surface, small droplets of water will form.

The dew point is the temperature at which moist vapor turns to liquid water. Air with high humidity will have a high dew point. That means that moisture will form on surfaces at higher temperatures. Examples of interior spaces that typically have higher humidity levels in cold temperatures are museums, enclosed swimming pools, and rooms with extensive computer equipment. Air with low humidity (dry air) has a low dew point. Moisture forms on the surfaces at lower temperature. A psychometric chart is a useful tool to determine the temperature at which condensation will form for a given humidity and dry bulb temperature.
The risk of condensation on glass can be greatly reduced with the use of insulating glass. Insulating glass consists of two or more panes of glass that are joined to the same structural frame (spacer), creating a sealed air space between them. The technology of double-pane glass was developed in 1934. The air space acts as a buffer and can be filled with dehydrated air or inert gas in order to improve its performance. The spacers in modern edge seals are typically aluminum and are provided with desiccant material, which absorb the moisture of the trapped cavity air. This desiccant reduces the dew point of the enclosed air to below -30° C (-35° F).

Provided these moisture absorbers are not damaged, they will continue to perform for up to 35 years. Once the drying capacity of the spacer is exhausted, fogging can occur. The seal, which joins the two sheets of glass and which is called a hermetic seal, consists of two seals. The first seal is made of butyl. Butyl has a very low perm rating and prevents moisture from entering the glass air space; however, it is not very strong and requires an outer seal of silicone or polysulfide, which prevents the two lights of glass from separating. When the hermetic seal fails, vapor penetrates the spacer and condensation forms on the inside surfaces of the insulated unit. Examples of failed insulating glass units can be seen on the south façade of the Institut du Monde Arabe in Paris.

A failed hermetic seal can be precipitated by the accumulation of water on the top surface of the insulating glass units. It is not uncommon for the weather seals of a curtainwall system to fail with age and a lack of maintenance. A secondary problem that can result from infiltrating water is the trapping of water in the head glazing rabbet. The glazing industry contends that trapped water in contact with a hermetic seal can cause problems with the glass unit. Insulating glass units can fail if the head glazing rabbet does not allow water to weep out of the system. A failed hermetic seal will appear as a fogged-up window that cannot be cleaned.

The airtight sealing of the cavity of an insulating unit is not a vacuum, and thus is susceptible to altitude changes. Optical and acoustical changes to the glass due to pressure differentials can occur if the glass is installed at a different altitude than where it was fabricated. For example, if the fabrication of the insulating glass lies several hundred meters above the place of installation, large panes might bow inward, actually coming in contact.

Ventilation

Many enclosure systems try to reduce or eliminate condensation through the use of mechanical ventilation. If warm air is blown on the glass, the temperature of the glass can be increased in order to prevent condensation. We are all familiar with how this works through the venting of the inside surface of our car windshield.

Completed in 1997, the four towers of the Bibliothèque Nationale stand like open books in the Paris skyline. The 20-storey towers are filled with books carefully protected from drying out by an extensive building humidification

right When pressurization fails, the windows will fog up. Bibliothèque Nationale, Paris

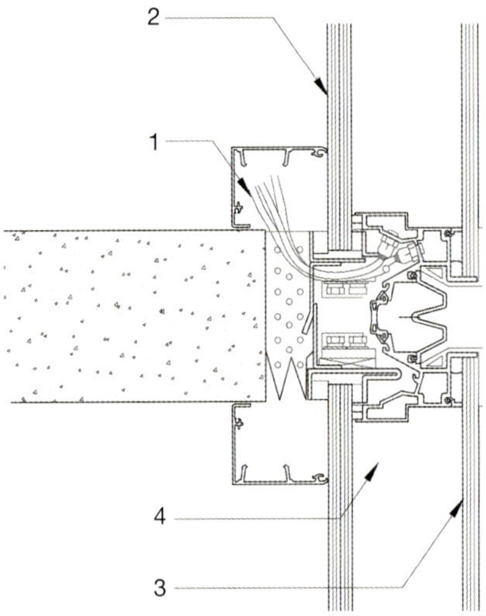

above Sill detail at the exterior wall
Bibliothèque Nationale, Paris:
1 Pressurization tubes connected to a compressor
2 Laminated fire-resistant interior light
3 Monolithic exterior glazing
4 Cavity temperature is controlled by pressurization.

system. Interior wood veneer floor-to-ceiling panels have been installed on pivots and provide protection from the sun. The 1.8 x 3.6 m (5 x 10 foot 5 inch) glass enclosure panels of the building limit the amount of ultraviolet radiation to the interior by means of laminated glass, and the interior pane adds another level of protection with fire-resistant glass. Extra-clear glass (low iron) was used to increase the sense of transparency. At the same time, the wall had to be designed to resist condensation from forming on interior surfaces that were exposed to high humidity levels all year round. The buildings humidity controls are critical to preserving the paper documents within. The Bibliothèque Nationale's specialized curtain-wall composition has an interesting method of combating the formation of condensation. Unlike traditional insulating glass units, which provide a hermetic seal around the airspace between the lights of glass, the Bibliothèque cavity is pressurized with individual compressors, which prevent condensation from accumulating within the glass cavity. The floor-to-floor glass façade has a wide cavity of 90 mm (3 1/2 inch), which is connected to a pressurized air system. The pressurized air is fed through valves in the lower frame profile and led out at the top near the vertical profile. Depending on requirements, the filtered air can be pre-heated, or cooled, to balance the temperature difference between outside and inside. In concept, this prevents condensation from forming on the glass. This is a critical factor for the high humidity levels of the building.

In locations where the system is not working, condensation forms quickly. One problem with this enclosure is that the design relies heavily on supplemental components. If a compressor fails or the piping to the cavity clogs, the glass will fog up with condensation. This sophisticated enclosure system is dependent on a multitude of compressors, seals, and tubing to maintain its goal of transparency. Because of its unique configuration and complexity, the windows can be difficult to repair. Not unlike the south wall of the Institut du Monde Arabe, the Bibliothèque Nationale's enclosure may meet with economic constraints by trying to maintain a truly courageous and innovative façade design.

When moist air comes in contact with cool surfaces, moisture appears as condensation. As outdoor temperatures decrease, condensation forms on the inside surface of monolithic exterior glazing. Condensation on monolithic glazing is a common problem for merchandise storefronts where the view of products for sale is impaired by this fogging effect. It can also be a problem for buildings that rely on maintaining vistas through the glass.

The original Reichstag building in Berlin was opened in 1894, destroyed by arson in 1933, and rebuilt. The reconstructed dome of brick,

right Acrylic louver blades move and rotate to follow the sun's path.
Siemens Pavilion, Seville

center Acrylic louver blades move and rotate to follow the sun's path.
Siemens Pavilion, Seville

below Glass louvers in section and plan; the technology inspired laminated glass:
1 Light is reflected back into the building by a specially designed lamination.

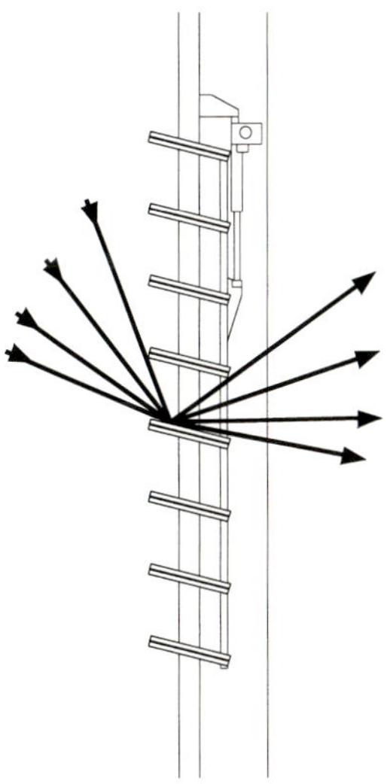

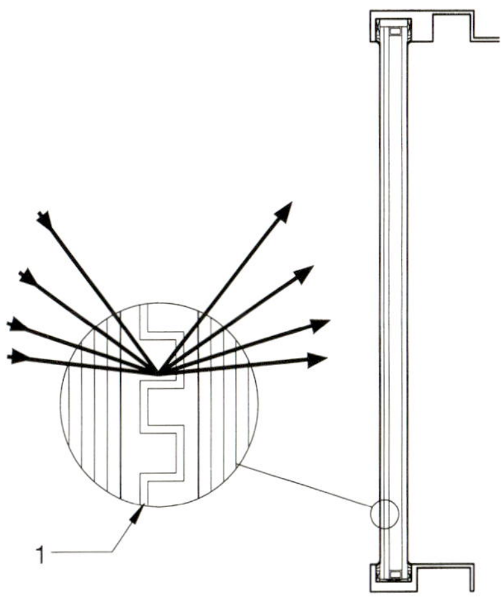

steel, copper and glass was demolished at the end of 1954 after sustaining heavy war damage. After German reunification in 1990, the German parliament decided to move back to Berlin. In 1992, a competition was held for remodeling the building. The quest was to transform the Reichstag building from a symbol of German history to a symbol of the future. Completed in 1998, the eventual design had a ribbed glass structure as opposed to the original dome. The dome was made accessible to the public. The steel framework of the dome, which has a diameter of 40 m (115 foot), supports the two walkways inside the

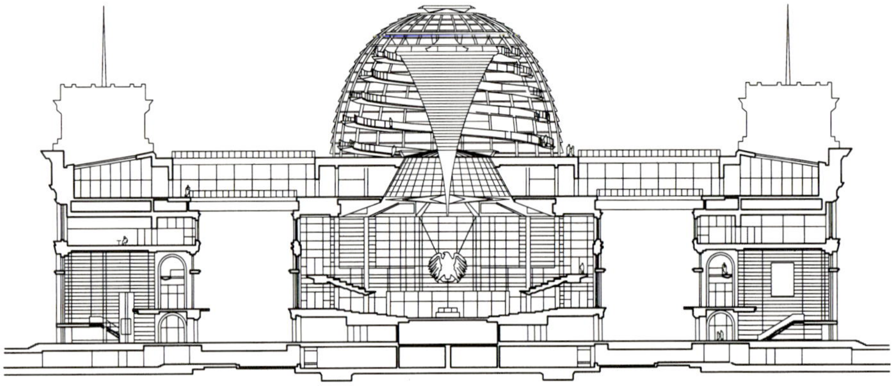

above Reichstag Dome, Berlin

center Section of Reichstag

right The bottom segments of the dome were not filled with condensation and appear to have natural ventilation across the inside surface of the monolithic glass skin. Reichstag Dome, Berlin

page 101 Reichstag Dome, Berlin

dome. At the center of the dome, there is a carrot-shaped cylinder with 360 individual mirrors providing light to the parliamentary chamber 10 m below. The cylinder also provides ventilation (fresh air) for the debating chamber. The air passes upwards via an air-extraction nozzle – by thermal up draught – and leaves the building through a central opening in the dome. The outer skin of the dome has an area of 3,000 square meter and consists of transparent glass segments. Clarity of structure, generosity and transparency are the special characteristics of the Reichstag. Visitors to the building enjoy a panoramic view of the adjacent Tiergarten and central Berlin from the walkway inside the dome. Transparency of the glass structure is critical to the design of the enclosure.

The dome is not an enclosed, environmentally controlled space, and therefore is not designed to be a weather-tight enclosure. As one enters for the first time, the openness of the structure is surprising. The dome is uncovered at the top and open at the bottom perimeter. A net at the top (barely visible) prevents birds from entering the dome. From the ground level, the dome provides an amazing surreal view at night as visitors ascend to the top. The dome skylight details are composed of a two-sided captured frame supported at quarter points with aluminum outriggers.

above Top of Reichstag Dome, Berlin

below Condensation can be seen on the monolithic enclosure.
Reichstag Dome, Berlin

Because of the variable humidity and temperature conditions in the space, the dome cladding design can collect condensation under certain environmental conditions. Although it is a rare occurrence, visitors to the dome at certain times of the year can be disappointed when a portion of the monolithic glass enclosure is fogged with condensation. The bottom rows of glass at the dome are open to allow for ventilation across the glass. These glass surfaces do not appear to accumulate condensation.

How can Glass Failures in Energy Conserving Systems be Avoided?

This is a very difficult question to answer. Our world has a drastic need for buildings to reduce their energy consumption. New technologies are constantly being developed to meet this need. With any new technology there are lessons learned through trial. This is a necessity of innovation, and a requirement in the development of even better solutions. For example, the development of the glass sunshades, which reflect sunlight back into the building, has inspired the development of a new type of laminated glass technology. The principle behind reflecting natural light back into our buildings as a method of saving energy is being developed at a microscopic level with a specially designed interlayer that accomplishes the same effect in a transparent light of laminated glass. These innovative laminated lights of glass reflect sunlight on the outer pane up toward the ceiling and further into the room. This new technology represents the next generation of glass enclosures that will bring more natural light into buildings.

The quest for energy conservation has not stopped with the complex systems developed in the 1980s. We are on the cusp of manufacturing a glass building material that could function as the building's exterior skin while simultaneously providing energy to the building. These glass panels, unlike the solar panels of the 1970s, allow for visible light to transfer through them (5 % light transmittance) and produce clean electricity from the sun's rays. The use of BIPV (building-integrated photovoltaics) is a serious effort toward sustainable design. The goal is to have our homes and businesses produce all, or nearly all, of their own electricity. If effective, the technology will make great strides in reducing greenhouse gases that harm the environment. Photovoltaic (PV) glass panels are a technology that converts sunlight directly into electricity. The technology is advancing closer to being used as a substitute for glass enclosures that form the outer envelope of our buildings. Integration of PV is basically a direct replacement of the normal glass panels with a PV laminate.

above Photovoltaic glass: BP Stations throughout the world are being designed with energy-producing glass.

right How do photovoltaics work?
1 The sun emits photons (light), which generates electricity when they strike a photovoltaic cell.
2 Both layers of a two-layer cell are made of a semi-conductive material, such as silicon. The layers are given opposite charges – one positive and one negative.
3 Electric current is conducted to an external circuit to power the appliance.

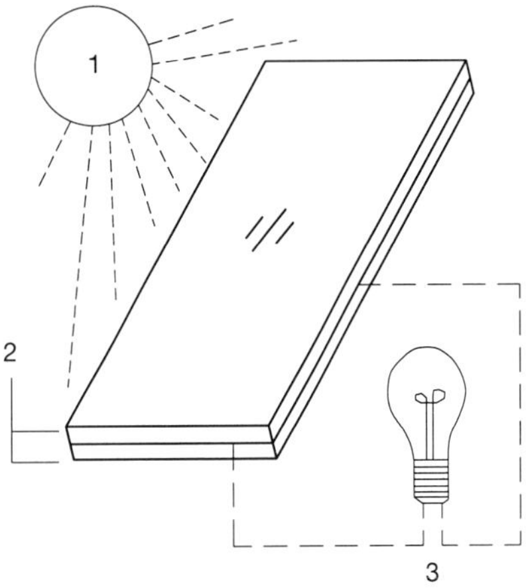

The goal is for all the electricity produced by the photovoltaic glass to be consumed within the building for things like lighting and HVAC power (heating, ventilation, air conditioning), but if there is excess of power, it can be fed into the local utility grid, generating a credit to the building owner.

Both of the innovations described above have limitations and are still being developed; however, they represent the next step in glass technology. In our search for more energy efficient enclosures and sustainable building designs, we will be faced with glass failures that are inevitable. It is important to understand the risks of newer technologies and not to lose sight of the goal.

Lessons Learned

1
Insulating glass units can prevent condensation and improve the thermal performance of an enclosure. Insulating units fail with time. When this happens, the visibility and color of the units will greatly change. The warranties are typically voided if the seals to the units are put in water. This can happen at the heads of the windows if water is not allowed to drain properly.
2
Sophisticated enclosure systems can be courageous architectural elements that enhance the appearance of building façades, but they add a higher level of maintenance and complexity to the building.
3
Double skin façades with internal shading devices are sophisticated enclosures of the future and have successfully been installed throughout Europe. Architects, contractors, and building owners must realize that initial costs may not be the only danger lurking over these systems. Double skin façades are susceptible to the same types of enclosure failures as traditional curtainwall systems. In fact, this type of enclosure may be more susceptible to failures because of its complexity and additional quantity of glazing.
4
Natural and mechanical ventilation is a proven method of reducing the risk of condensation forming on glass. Both of these methods have limitations that must be considered in designing enclosure systems.
5
In the quest for conserving energy, innovative enclosure systems will challenge traditional methods of façade construction. This is an important part of the development of architectural façades. These innovations will also be challenged by problems and criticism. By understanding the limitations and benefits of these new systems, we can strive toward further improvements

Structure

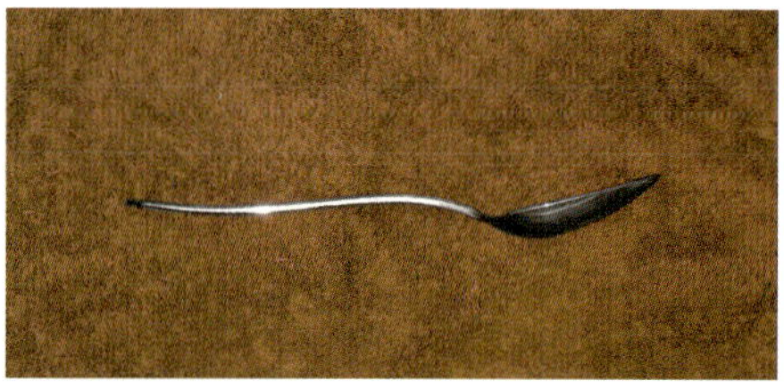

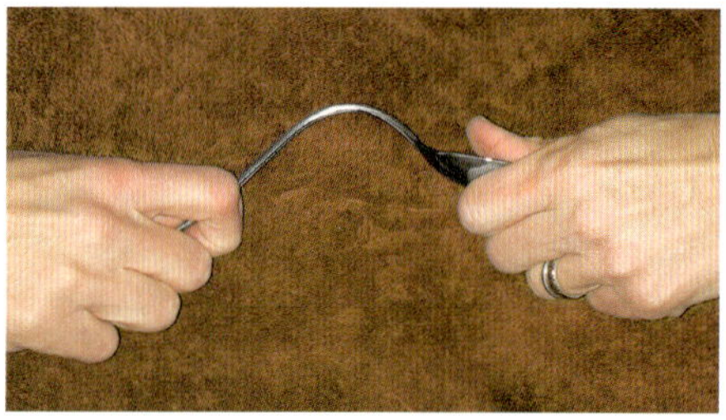

The unique and important quality of glass is its ability to transmit light. Unlike other solid construction materials, glass provides the transparency which has become a necessity for virtually all buildings. With glass's unique transparent nature comes an equally distinct brittle behavior that is uncharacteristic of traditional building materials. Unlike metal or wood, glass exhibits brittle properties under pressure. This means that at the point of failure glass does not yield, but breaks in an abrupt manner. Steel and wood members can redistribute load or absorb energy by yielding, which makes their behavior ductile. We understand the ductile nature of metal from the simple exercise of bending a spoon. After exerting a certain amount of force and releasing, the pencil maintains a portion of the deformed arch as its permanent shape. This is the characteristic of a material with some plastic deformation.

Glass is a liquid that is cooled to a solid state without crystallizing. Some classify glass as a super cool liquid; however, this suggests that glass still flows. Glass is rigid and will not flow in normal temperatures. Many people falsely believe that glass continues to flow in its cold stage after fabrication. This myth is supported by examining glass in older buildings and measuring the thicker glass at the bottom of the pane. It appears that the glass continues to flow in a fashion similar to a slow moving liquid. In truth, the thicker material is due to early glass manufacturing techniques and does not represent a transformation of the glass form. The thicker material could be just as likely found at the top of the glass window.

Glass is not ductile. A glass stirring rod will bend under an applied force, but will bounce back to its original position once the force is released. The brittle nature of glass will not allow for the redistribution of loads or the absorption of impact energy. It will not permanently deform under stress. Bent glass is produced not by the glass yielding under force but by heating annealed glass to the point

right and page 105 Completed in 1999, this U-Bahn station represents the lighter structural forms of newer buildings.
U-Bahn Station St.-Quirin-Platz, Munich

where it softens, so that it can be pressed over forms to give it shape. If extreme forces are applied, the glass will shatter. In an instant, the material's ability to resist the force is gone. The source of virtually all glass breakage is tension. It is not possible even to measure the compressive strength of glass, because the tested sample will fail due to the tension stress side effects in the glass. Most tension failures in glass originate from defects in the glass surface. Glass has overcome its fragile nature to become a load-carrying element in our contemporary building enclosures. Today, glass is used to resist design loads more commonly supported by steel, wood, and concrete members.

Glass can be very strong, much stronger than steel; however, because glass has a relatively low fracture toughness, its highest strengths can only be achieved when it is free of defects. Glass technology has challenged its susceptibility to tension failures with the development of the heat-treating process. Tempering, also called toughening, allows for greater strength properties by putting the outer layer

right Structural weakness in enclosure systems can be uncovered through mock-up testing.

page 107 Bent glass is produced not by the glass yielding under force but by heating annealed glass to the point where it softens, so that it can be pressed over forms to give it shape.
Ferriswheel "London Eye"

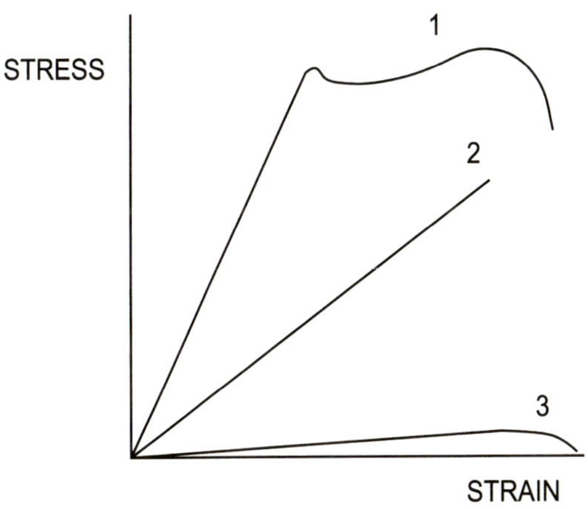

Stress-strain diagram:
1 Steel
2 Glass
3 Wood
The graph describes stretching or compression of a material under an applied load. The stress-strain curve of glass depicts its brittle nature.

of glass in compression. This added strength comes with some limitations that should not be overlooked. Tempered glass allows us to use glass in structural applications with greater confidence. This confidence can quickly be shattered if details and design loads do not consider the inherent fragile nature of this unique material.

Completed in 1998, the Palais de Justice building in Bordeaux used glass as a structural element for its enclosure system. The building design program called for seven courtrooms, administrative offices, and a large public space leading to the courtrooms. The French Ministry of Justice requested a building that would emphasize the accessibility of the French judicial system to the public. The design of their new building achieved this by providing the feeling of transparency and openness around the courtrooms. The seven courtrooms are pod-shaped timber-clad volumes, supported by concrete pilotis. The wooden courtrooms combine high-tech geometry with traditional construction materials.

Recent developments in computerized machining were essential to the production of the complex curves and acoustic perforation of the internal maple-veneered finishes. The seven pod-shaped courtrooms stand behind a nearly invisible glass curtainwall. Vertical glass fins were designed to resist the forces of wind and rain on the glass.

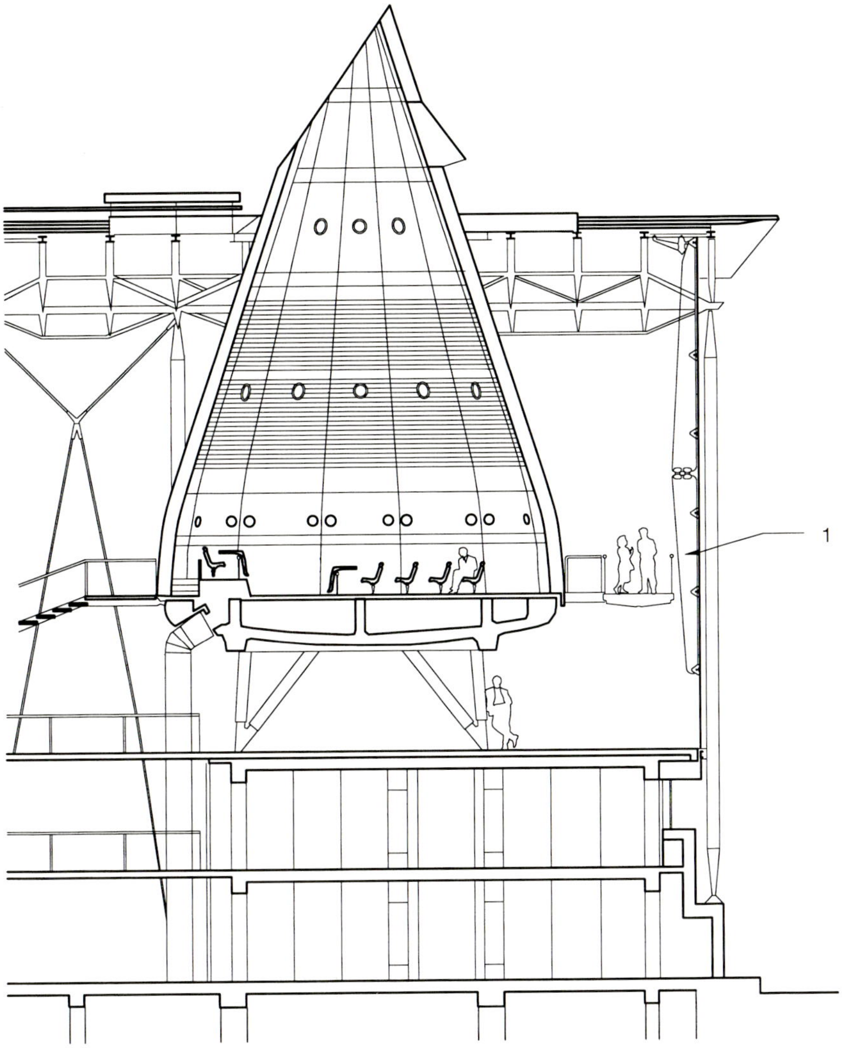

above The pod courtrooms are visible to the public through a transparent glass wall supported by glass fins.
1 The glass fins were replaced with steel fins due to breakage.
Palais de Justice, Bordeaux

right The glass fins on the all glass façades were wrapped in white canvas as a safety precaution prior to replacement.
Palais de Justice, Bordeaux

below The replacement fin is a less transparent but more ductile steel plate.
Palais de Justice, Bordeaux

page 108 The steel replacement fins are left exposed.
Palais de Justice, Bordeaux

Shortly after completion, the exterior enclosure began to have problems. The vertical glass fins which support the lobby glazing started to fracture. The glass fins were wrapped with canvas tarps to protect visitors from breaking glass. The cause of the glass failure may have several contributing factors. Finally, a steel fin replacement was developed. The original glass details were custom-made for the project; however, similarly designed glass fin details had been successfully installed on numerous other projects. The challenge with building this type of enclosure is executing details that flawlessly transfer forces through the system. The connection between glass and metal is crucial to that transfer. Typically, the transfer connection includes a steel bolt that penetrates the tempered glass. A liner material is used to separate the hard metal from the glass. The liner can be made of a variety of materials from aluminum to neoprene. If aluminum is used, it is important to install a soft aluminum alloy that will evenly transfer the large forces to the glass. If a neoprene material is used, it should not be so soft as to displace under extreme pressures. The new Palais de Justice building is a marvelous design, which was unfortunately tainted by the failure of the glass fin support system. The project compels us to realize how sensitive and non-ductile glass can be. In order to better understand how glass is used as a structural element, a fundamental understanding of its mechanical properties should be reviewed.

Strength Properties of Glass

The properties of glass are well known but the strength of glass has a design parameter not common to other building materials: breakage probability. Unlike other building materials, glass strength is inherently unpredictable. The strength of two identical pieces of float glass tested under identical conditions can vary by a factor of three. Unfortunately, it is not possible to predict the strength of any one particular piece of glass without testing and breaking it. In 1921, A. A. Griffith developed the theory that the surface of a piece of glass is filled with microscopic defects (Griffith Flaws). Because glass cannot plastically deform and redistribute applied loads, it is susceptible to localized over-stressing and is very vulnerable to flaws in the surface of the material. The strength variability of glass is directly dependent upon such things as cracks, defects, flaws, and other characteristics related to that piece of glass. Any one of these defects, under certain loading conditions, may concentrate the applied stress on to the interatomic bonds in the defect and cause the glass to break.

Empirical glass strength curves have been developed from the testing of 2,000 pieces of glass. This data helps to predict with reasonable confidence the stress levels that glass can sustain; yet, because of the variability of glass, it is very difficult to achieve 100 % assurance that all specimens will act in the same way. This leads us to a familiar term in the glass industry: breakage probability. The probability that a glass light will break under a certain loading can be estimated using experimental data. Glass in enclosure systems is typically designed to meet a 8/1000-breakage probability. This means that under the design stress eight lights in 1,000 may break. If we would like to reduce the risk of glass breakage to perhaps 1/1000, we will be forced to use a smaller design strength and possibly be required to increase the thickness of the glass.

Computer techniques utilizing finite element analysis were developed in the 1970s, which made it possible to determine the structural performance of float glass under uniform wind loading. Glass manufacturers and suppliers have taken on the responsibility of determining the glass thickness required for a particular application.

Mechanical Properties of Glass

E Modulus of Elasticity (Young's Modulus)	10.4×10^6 psi 7.2×10^4 N/mm^2
G Modulus of Rigidity (Shear Modulus)	4.3×10^6 psi 3.0×10^4 N/mm^2
Poisson's Ratio	0.22
Coefficient of Thermal Expansion	49×10^{-7}/F° 88×10^{-7}/C°
Density	157 lb/ft^3 2.5 g/cm^3
Modulus of Rupture (Flexure)*	
Annealed	6,000 psi 43N/mm^2
Heat-Strengthened	12,000 psi 82.7N/mm^2
Fully Tempered	24,000 psi 165 N/mm^2

* Modulus of rupture based on a 60-second load

center Fracture patterns:
1 Annealed glass (normal glass)
2 Heat-strengthened glass
3 Tempered glass

below Differential pressure: If the air pressure between the lights of glass of an insulating glass unit is significantly lower than the outside air pressure, an optical distortion will be visible on the glass surface.

Heat Treatment of Glass

In theory the compressive strength of glass is around 21,000 N/mm^2 (3,000,000 psi).Unfortunately, it has never been measured. Testing of glass in compression invariably produces tensile stresses as a side effect, which breaks the glass. Glass for all practical purposes never fails in compression but rather from tensile stresses. Normal window glass usually fails at stress levels less than 100 N/mm^2 (14,500 psi). Because surface flaws lead to fracture only when a tensile stress opens them, any method that keeps the glass in compression is advantageous. Tempered and heat-strengthened glass use this principle. These heat-treating methods increase the strength of annealed glass by providing residual compression stresses in the surface of the material. The tempering process leaves the outer layer of glass in compression with the inner layer in tension. If any crack penetrates the outer compression shell of tempered glass, the residual tensile stressing of the core will shatter the material. One can experience this effect by inducing an acute force, for example from the tip of a nail, into the edge of a tempered light of glass. The result will be the explosive propagation of the entire sheet into a pile of tiny particles.

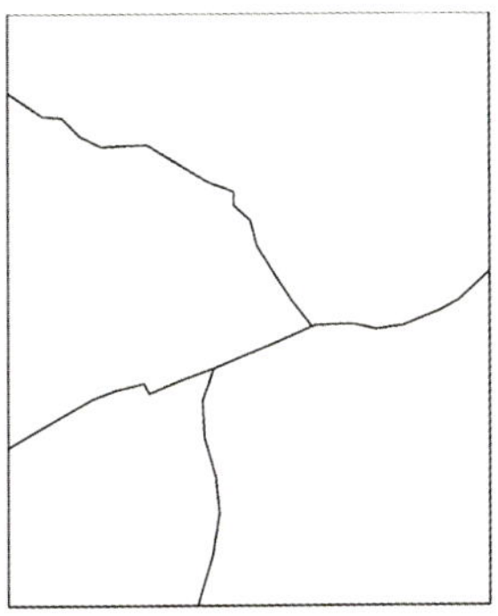

1

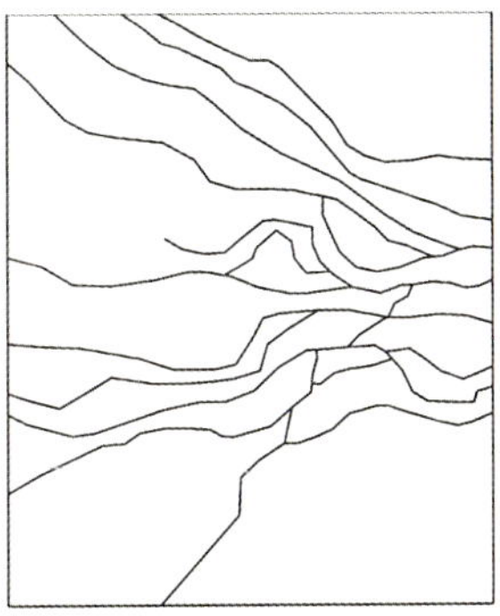

2

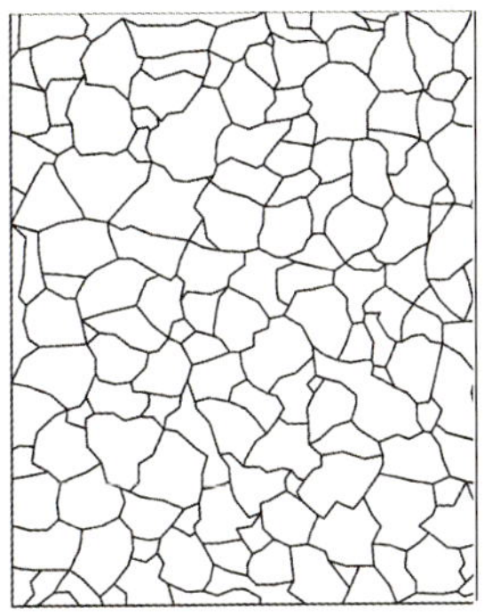

3

When annealed glass is heat-treated, it is put through a special process involving heating and cooling. The heat treatment process does not affect the glass clarity, expansion properties, surface hardness, color, light, or heat transmission properties. All of these properties remain virtually the same. The strength of the glass, however, is greatly increased. Tempered glass is virtually four times as strong as annealed glass.

Heat-strengthened glass typically has a surface compression between 3,500 and 10,000 psi (24 and 69 N/mm^2). Tempered glass has a surface compression of 10,000 psi or greater. At its upper limit, heat-strengthened glass has strength characteristics similar to those of fully tempered glass. In many cases the breakage characteristics for upper-level heat-strengthened glass are similar to breakage characteristics of tempered glass.

This added glass strength does come with minor limitations. Fabrication and cutting of tempered and heat-strengthened glass must be completed prior to heat-treating. Any attempts to cut, drill, grind, or sand-blast the glass after tempering will result in glass breakage. Wired and deeply patterned glass cannot be heat-treated. Tempered glass is susceptible to nickel sulfide inclusions and spontaneous breakage. Finally, tempered glass also has a reputation of having minor visual distortions.

Roller-Wave Distortion

A minor side effect to the tempering process is roller-wave distortion. When annealed glass is tempered, it is sent through a furnace on rollers, heated to extreme temperatures and then cooled very quickly. Unlike the float glass process, the tempered glass is cooled on a roller bed, which leaves a slightly wavy appearance on the glass. The distortion is more pronounced on projects where untreated glass is adjacent to tempered glass. The Farnsworth House in Plano, Illinois, was constructed in 1950 with large lights of flat annealed plate glass. As a result of flooding from the adjacent Fox River in 1997, one of the large lights of glass was broken and subsequently replaced with tempered glass. The slightly wavy nature of the tempered replacement glass is detectable against the original untreated glass panes. Roller-wave distortion should not be confused with the optical irregularity of glass bending under differential pressure. If the pressure between the lights of an insulating glass unit are much higher or much lower than the surrounding atmosphere, the glass can balloon outward or cave inward causing optical distortion. This problem can occur when insulating glass units are made with large, thin lights of glass that are fabricated at a different elevation than where they are installed.

page 112 above The original design included glass sheets that were 3,450 mm wide by 5,380 mm tall.
Neue Nationalgalerie, Berlin

page 112 left The tempered replacement panels are not as flat as the original glazing. Roller-wave distortion creates a slightly wavy appearance to the glass if viewed from certain angles.
Farnsworth House, Plano, Illinois

page 112 below Replacement glass of the same size as the original glazing is difficult to obtain. The museum had to install two lights of glass separated by a sealant joint.
Neue Nationalgalerie, Berlin

below Current glass fabrication sizes are smaller than what was required at the museum. The new glass is half the size (1) of the original panels, and a sealant joint (2) separates the replacement lights of glass.
Neue Nationalgalerie, Berlin

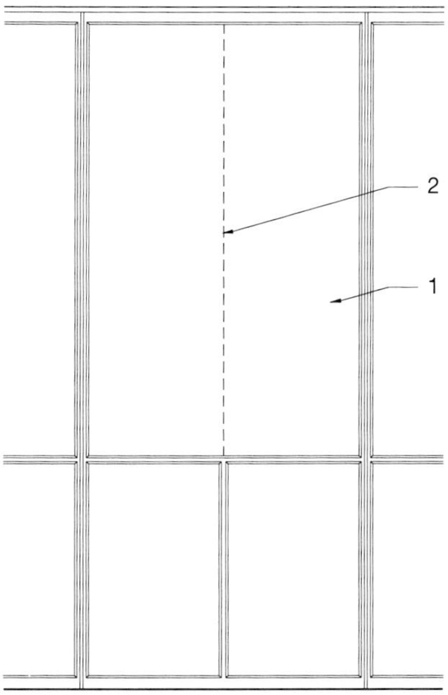

Strain Pattern Characteristics

In the production of fully tempered and heat-strengthened glass, the glass is heated close to its softening point and then quenched (cooled) with high-velocity blasts of air. This process inherently results in a slightly higher level of surface pressure on the glass directly opposite the air nozzles. This area will produce a slightly more dense glass material. As a result of this, the glass may exhibit a darker appearance under some viewing conditions. As the sun approaches the horizon, the natural light is polarized, and will highlight the dense glass areas. The perceived dark areas are the visual effects of the strain pattern. The intensity of these dark areas can be visible in thicker glass lights (thicker than 1/2 inch or 13 mm) with reflective coatings, and are not color distortions but inherent by-products of the manufacturing process.

Size Limitations

Heat-treated glass is limited by the width of the furnaces required to fabricate the material and the width limitations of the float glass industry. Because of the enormous expense involved with building the equipment, glass fabricators have geared their production to glass sizes that are most commonly used today. A casualty of this economic decision has been older glass enclosures made of exceptionally large lights of glass. Polished plate glass used to be made in large sheets that are wider than the contemporary float glass furnaces. Many completed projects struggle to find replacement glass for enclosures completed prior to the new methods of fabrication. The Neue Nationalgalerie in Berlin completed in 1968 had an enclosure construction that included glass sheets that were 3,450 mm wide (135.8 inch) by 5,380 mm (211 inch) tall. Today's float glass process typically limits the width of sheets to 130 inch (3,302 mm). As a result, replacement glass for the large openings on the building had to be installed using two individual pieces of glass with a sealant joint between them. Clear annealed glass of sizes as large as 6,000 x 3,660 mm are produced in certain parts of the world; however, they are not readily available to all building owners.

Laminated Glass

It is a common misconception that laminated glass, is stronger than annealed glass. In truth, laminated glass can be weaker than annealed glass, depending on the conditions to which the glass is exposed to. Laminated glass is produced by bonding two or more panes of glass together with a plastic material, a resin or intumescent interlayer. Laminated glass has many benefits. If it is broken, the pieces hold together. Laminated glass can incorporate other materials, such as polycarbonates, to achieve specific mechanical properties (i.e. bullet resistance). The laminating process affords the glass industry with a multitude of sound-reducing glass products, which combine a variety of glass materials to reduce the amount of sound that enters a building.

Laminated glass should not be considered to have the same load resistance as monolithic glass of the same thickness for sustained loads because of the shear creep of the plastic interlayer. An example of a sustained load would be the pressure from water in an aquarium tank on a pane of laminated glass. Typically, the interlayer is stiff enough not to allow the individual panes of a laminated piece of glass to slip over each other under short duration loads like wind pressure. However, interlayer material tends to creep under long duration loads. If this type of loading is anticipated, it is necessary to assume that the individual panes of glass can slide past one another and share the load between them. Thus, a laminated glass consisting of two 15 mm (5/8 inch) glass panes would be assumed to take approximately half the load on each pane. The effect on load capacity is that two laminated 15 mm (5/8 inch) glass panes are equivalent to a 21 mm (7/8 inch) monolithic piece of glass (not a 30 mm or 1 1/8 inch). If three 10 mm (3/8 inch) glasses are used in a laminated unit, the effect is worse. The strength of the triple ply 30 mm (1 1/8 inch) thick unit would be equivalent to a 17 mm (11/16 inch) monolithic piece of glass.

right The exterior façade is composed of insulating glass and supported by four through-bolt fasteners. Marché Saint-Honoré, Paris

center The building is placed on a site formerly occupied by a parking garage. Marché Saint-Honoré, Paris

below Laminated glass is susceptible to the same types of glass breakage as monolithic glass. Marché Saint-Honoré, Paris

Research has also determined that laminated glass behaves less like monolithic glass and more like layered glass as the temperature of laminated glass increases from normal conditions (below 100° F or 38° C) to elevated conditions (over 120° F or 49° C). This adjustment

in behavior is accompanied by a loss in glass strength and deflection characteristics. The properties of laminated glass are sensitive to temperature, because of the connecting material. At high temperatures, (over 170° F or 77° C), the adhesive of the interlayer is ineffective and the layers act independently. At freezing temperatures, the interlayer is very effective connecting the lights of glass and all layers act as a single unit.

The Marché Saint-Honoré Office Building is adjacent to Place Vendôme (famous for jewelry stores). Located in the center of Place du Marché Saint-Honoré, the site completely occupies an area formerly used as a parking garage. The building is encapsulated by an innovative double skin façade composed of an exterior insulating glass unit (laminated on the inside light of glass) with point-supported lights connected to a monolithic interior light of glass. Completed in 1999, the building has had a history of glass failures, which have continued to plague the façade since the building's opening.

A review of the building in November of 1999 indicated that the failures have occurred on all four of the sides of the building and was found to occur on the large lights of glass on the lower part of the building. The failures appeared to occur on the interior laminated light of the insulating glass unit. Point-supported lights of glass are extremely susceptible to high concentrated loads, which can lead to failures. Because the glass is laminated, it does not displace. This is an added benefit to the building's security; however, the project serves to remind us that laminated glass is still susceptible to the same type of glass failures as monolithic glass.

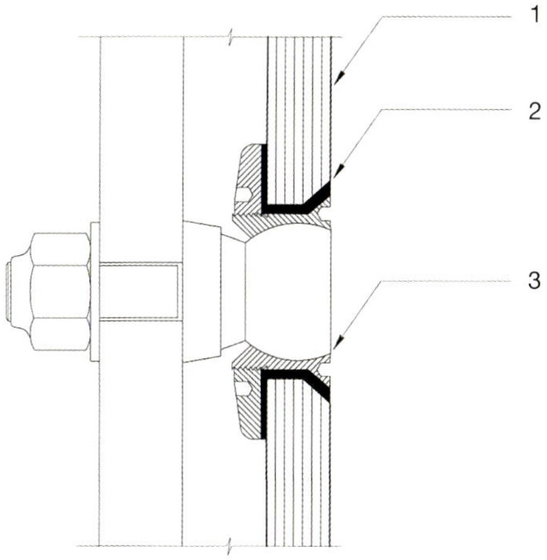

Glass isolation
Glass must be separated from hard surfaces by a softliner
1 Tempered glass required at point fixings because of high stress levels on glass

2 Liners can be rubber or a soft aluminum alloy. A hard aluminum alloy liner can cause glass failure.
3 The steel connection allows rotation of the glass panel.

above The original glazing system had a history of glass breakage which required a complete replacement of the glass façades.
The College Life Insurance Building, Indianapolis

right The replacement enclosure system's mullion spacing and framing composition is completely different from the original design.
The College Life Insurance Building, Indianapolis

Isolation of Glass

Whether glass is annealed or heat-treated, it is susceptible to acute loads and cannot accommodate local over-stressing. Therefore, all materials in contact with glass should be relatively soft and distribute forces to and from the glass. This concept explains the common glazing practice of avoiding metal-to-glass contact. We are all familiar with the rubber gaskets that surround the windows on our cars. In addition to stopping air and water, these materials are used to distribute forces imposed on the glass into the surrounding frame material. This transfer of load is done without exposing the glass to acute localized pressures. Glass in windows typically rests on setting blocks. These blocks are made of a forgiving material like neoprene and spread out over several inches to transfer the weight of the glass to the frame below it. Glass can also be fixed to a supporting frame by using bolt fixings and spreader plates. The use of these types of connections will invariably result in localized stress concentrations around the connection. For this reason, tempered glass is typically required because of its higher strength properties. The liner or bushing, which separates a metal bolt from the tempered glass, must be hard enough to prevent excessive deformation yet soft enough to distribute forces evenly around the glass.

Failure of Glass Support

Glass breakage in buildings is not always caused by a failure in the glass material. More often than not, the material used to support the glass is the weak link in the enclosure. If the framing system has not been adequately designed to resist the forces on the building, the glass will break as a result of the deformation of the frame that supports it. The College Life Insurance Building in Indianapolis, Indiana, required re-cladding of its entire façade after the glass support system was found to be inadequate. Completed in 1972, the three 11-storey office buildings were the initial phase of a much larger development project that was never realized. Their distinctive form represented a new landmark for the city and was celebrated in architecture magazines and history books. Each of the three towers is composed of two walls of glass and two walls of concrete. The glass walls are made of a structural neoprene sloped glazing. Approximately 50 % of the 4,000 windows in the pyramid-shaped buildings developed serious problems shortly after the building was occupied. The problem involved cracks in the 4 x 6 foot (1.39 x 2.09 m) insulating glass windows, followed by fogging of the inside surface of the double-glazed units. Water leakage into the interior was excessive at the windows. Despite attempts to correct the problems with new

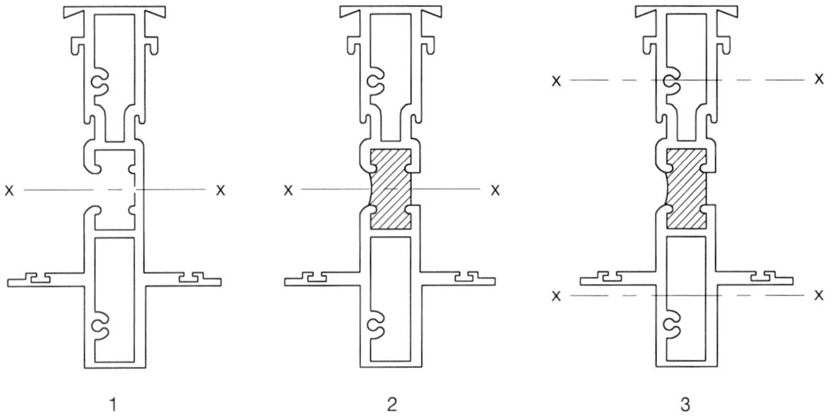

above Poured and debridged extrusions will have a fraction of their structural capacity if the urethane separator is compromised:
1 Prior to debridging
I = 1.6 inch^4 (67 cm^4)
2 After poured urethane becomes rigid and aluminum connection is removed, the stiffness is the same I = 1.6 inch^4 (67 cm^4), but the thermal performance is greatly improved.
3 If urethane bridge becomes soft or displaced, the combined I of the two sections = 0.288 inch^4 (12 cm^4), a fraction of its original stiffness.
(I = moment of inertia)

window gaskets and new glass, the leakage and glass failures continued. Field testing of the existing system after problems had started indicated that the mullions could only support a fraction of the design loads. In the end, the entire curtainwall system was replaced. The replacement system has a completely different glass module and supporting mullion structure. The only component of the original system is the glass. Because there was no problem with the original glass design, the same type of glass was used for the replacement system.

Thermal Break Problems

Sometimes structural problems in glazing systems can be the result of the change in temperature. A thermal break is a technique used in a window system to isolate the outside metal from the inside metal. The detail is used to conserve energy and reduce the risk of condensation forming on the inside surface of the enclosure. There are many types of thermal breaks. The use of poured and debridged polyurethane thermal barriers is one of the most common methods of achieving thermal separation in aluminum enclosures. A two-part high-density polyurethane is mixed as a liquid and poured into a dog-bone-shaped cavity of an aluminum extrusion. After the urethane cures, the bridge is milled away, leaving interior and exterior metal separated by a minimally conductive urethane material. The aluminum and urethane materials act in composite action to structural loads. The bond between the thermal break material and the aluminum extrusion is critical to the structural performance of the member.

Urethane expands and contracts approximately five times as much as aluminum. Therefore, the urethane will try to expand and contract more than the aluminum during thermal cycling. As long as the urethane adheres to the aluminum, the expansion and contraction will occur with no separation between the two materials. Dry shrinkage occurs when adhesion is lost and the end of the urethane separator shrinks back from the cut edges of the extrusion. This leaves a gap between the interior and the exterior aluminum. If this happens, the poured and debridged thermal breaks will not perform well after thermal cycling. Thermal cycling consists of exposing the material to low temperatures followed by high temperatures, then followed by low temperatures for a designated number of cycles. Exposure to an extreme temperature range can cause the thermal isolator to "de-bond" from the aluminum extrusions. If the materials act independently in a non-composite manner, their structural capacity will be greatly reduced. This reduction in strength is normally indicated by excessive deflection during wind loading.

Dry shrinkage can be avoided by improving the adhesion between the urethane and the aluminum or by restricting the amount of thermal expansion of the urethane. Adhesion can be improved by applying adhesive-promoting primers to the aluminum extrusions prior to filling and debridging. The thermal expansion of the aluminum can be restricted by placing an accordion-shaped aluminum bar in the cavity before filling it. This is similar to the way steel reinforcement in concrete prevents shrinkage; after the urethane cures, the bar dramatically reduces the amount of movement in the urethane under thermal cycling. Another method of reducing the movement of the urethane is by "knurling" the extrusion at 3 inch (76 mm) intervals the prior to installing the thermal isolator. The dimples created by the "knurling" improve the bond between the isolator and the extrusion and restrict the movement of the urethane during thermal cycling. In a testing environment, it is always important to provide structural loading of this type of system after exposing the material to thermal cycling. A mock-up test in a laboratory can uncover this problem prior to installation of the enclosure on the building.

Urethane thermal break problems are not limited to dry shrinkage. A building at the University of Wisconsin in Platteville had an entire section of the aluminum frame falling off the building in 1970 when the thermal barrier completely failed. Urethane samples from the project were analyzed and found to be grossly off ratio and to have never cured correctly. Today's quality control measures for urethane mixing makes the possibility of this occurring again unlikely; however, any change in the formulation of the urethane thermal break should be met with scrutiny.

Fastener Failures

Although usually concealed from view, enclosure anchors are a critical component to the performance of a system. The structural integrity of building skins is ultimately dependent upon the fasteners used to hold it together and connect it back to the building. Fasteners must be designed to include structural adequacy, adjustability, corrosion prevention, as well as to accommodate differential movements. They are also critical in accommodating building variations, erection accuracy, and alignment. The anchor system must conform to sometimes widely varying substrate conditions. The design of connections to resist the loads imposed on the building should be supported by sound engineering principles. Anchors must accommodate forces and thermal movements without failing. Nonetheless, unforeseen defects in the fasteners can lead to serious structural problems with the enclosure.

Hydrogen embrittlement is a condition of low ductility in metals resulting from the absorption of hydrogen during the manufacturing process. High strength bolts and screws that are exposed to hydrogen can fail prematurely at a stress level well below the material's yield strength. The hazard caused by hydrogen embrittlement is the unpredictable failure which may occur to a fastener under high tensile load. Electroplating is generally considered to be a cause of hydrogen absorption in steel fasteners, due to the release of hydrogen during the manufacturing process. Results of such failures can be disastrous. Early discovery of this type of problem will allow for replacement of fasteners prior to completing an enclosure installation. Higher strength steels are more susceptible to hydrogen embrittlement than lower strength steels; how-

Plywood panels fill the openings where the glass panels broke.
John Hancock Tower, Boston

ever, all steels are susceptible. Tests such as the incremental load hydrogen embrittlement test can be conducted to assess whether hydrogen embrittlement is present in a batch of fasteners.

Preventing galvanic corrosion is critical in fastener selection. Dissimilar metals in contact with each other in a wetted environment can lead to the corrosion of the least noble material. Some metals – for example, aluminum – in contact with concrete will corrode if exposed to moisture. This type of degradation to a fastener can loosen the connection and lead to failure.

Defective fasteners are not the only cause of connection failures for enclosure systems. Building envelopes are exposed to cyclic loading 24 hours a day. Winds blowing on and off of the building create an infrequent irregular loading pattern. The constant vibration of this type of loading can undermine the enclosure's connections. One mode of connection failure due to vibration is through the loosening of bolts that connect the enclosure back to the building structure. Bolted connections can be protected from loosening with the use of lockable fasteners. Fasteners can be locked to prevent loosening by several methods, including using specially designed locking nuts, washers, or peening the threads of the bolt after installation. A locked fastener will not loosen under cyclic loading. Understanding the types of loads that building skins are exposed to is an important step to preventing failures.

Fatigue Failure

As indicated by the stress strain diagram, glass deforms in a linear fashion. Reversal of stress leaves the glass unchanged. When a force is removed, glass returns to its original shape. Glass is not susceptible to dynamic fatigue from cyclic loading. However, the fatigue of materials which support the glass can result in glass failures. Fatigue can be defined as the tendency to fracture under cyclic stresses. Enclosure systems are exposed to constant variations in pressures.

The replacement panels are monolithic, fully tempered glass.
John Hancock Tower, Boston

The ever-changing positive and negative forces from wind make façade elements susceptible to fatigue failure.

A building enclosure that suffered from the effects of fatigue was the John Hancock Tower in Boston. Completed in 1973, the glass tower soared over its rustic masonry neighbors as a powerful symbol of modern architectural design. The project was nearing the end of construction when, on the night of January 20, 1973, dozens of glass units failed as a result of a storm. In the weeks that followed, about a third of the Hancock's glass was removed and replaced with plywood. Many of the broken glass units had been damaged by falling glass from above; however, the cause of the initial glass failure was unknown.

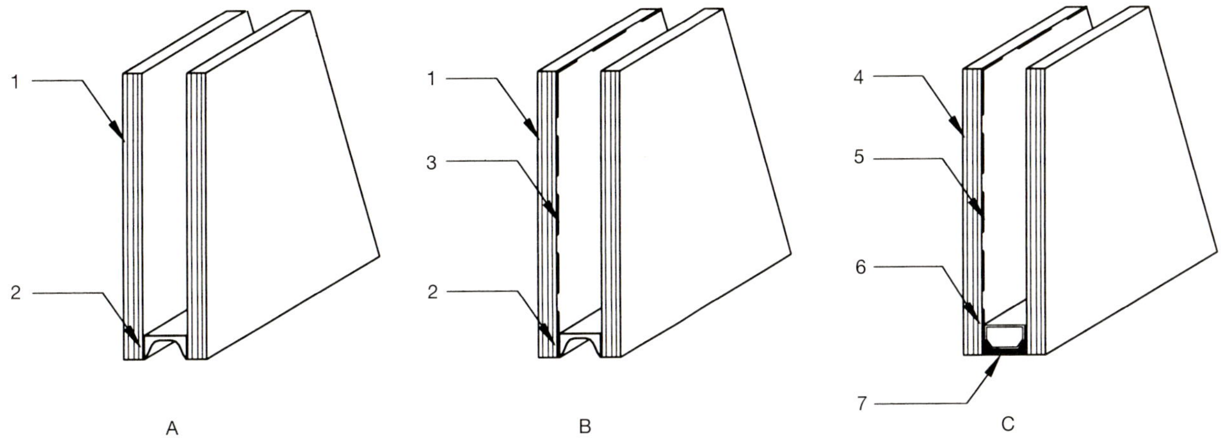

above Diagrammatic sketches showing the different compositions of insulating glass units:
A Insulating glass units prior to the John Hancock Tower
B Insulating glass at the John Hancock Tower
C Insulating glass construction today

1 Exterior float glass
2 Continuous lead spacer soldered to the exterior and interior glass
3 Reflective coating developed just prior to construction of the Hancock
4 Exterior float glass
5 Reflective coatings commonly used in contemporary insulating glass units
6 Metal spacer filled with desiccant and sealed to glass with butyl sealant
7 Polysulphide seal (silicone for most structural glazing applications)

A feverish search was started to find the cause of the problem. The Hancock was clad with over 10,300 glass windows that contained virtually the same glazing details. This unit consisted of a double-glazed window measuring approximately 4 1/2 x 11 1/2 foot (1.56 x 4 m). The size of the units was determined to be one of the contributing factors to the failures. The glass went from floor to floor and consisted of larger lights than typically used for this type of application. Prior to the building's construction, there had been fire code requirements for a spandrel between the floors with insulation to prevent flames from going up and breaking the glass on the floor above.

The Hancock was equipped with sprinklers and other life-safety advances; therefore, spandrels were no longer required. The outer lights of the insulating glass units were covered with a reflective coating to limit solar gain and provide greater thermal performance. The windows were made of annealed glass to preserve the clarity of the reflective glass. Heat-treated glass would have been stronger but would have had some roller-wave distortion due to the heat process.

below Diagrammatic sketches of wind study indicate that wind speeds (and equivalent one-minute pressures) do not necessarily increase with the height of the building and that the building edges have the highest wind loads.
1 10 psf (0.50 kPa)
2 12.5 psf (0.60 kPa)
3 17.5 psf (0.84 kPa)
4 25 psf (1.2 kPa)
5 32 psf (1.5 kPa)
John Hancock Tower, Boston

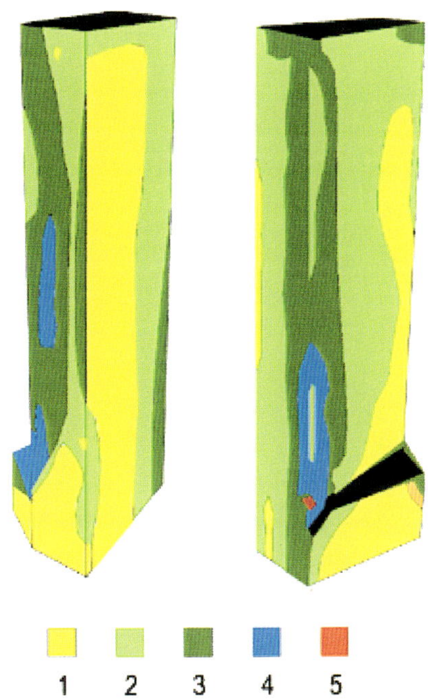

Wind tunnel study of CN Tower, Toronto, 553 m high and the world's tallest free-standing structure (1976).

The glass failure at the Hancock was extensively studied by Thomas Schwartz of Simpson, Gupertz & Heger, Inc. Schwartz has attributed the primary fault of the windows to the fabrication methods of the double pane and the reflective coating on the glass. Double glazing units had been around since the 1930s, and the Hancock was constructed using the same details common to the industry at that time. The insulating glass units that made up the building skin were fabricated with a thin lead tape spacer to separate the two panes of glass. After the glass was coated with a film of copper along its edge, the lead spacer was soldered with hot lead to both the front and rear pieces of glass to form the sealed unit. The thin film of copper on the glass insured a strong chemical bond to the lead spacer with the solder. The air space between the two lights of glass provided the thermal barrier for the window.

What made the Hancock glass unit different was that the inside surface of the outside piece of glass had a reflective coating applied to it. This reflective glass was a new architectural material; it started to become available in the 1960s. The glass coating was a thin layer of silver-colored material including chrome. This was an exciting material for the time, not only because of its appearance but also because of its improved thermal performance. The coating rejected some light, and the glass absorbed a considerable amount of the heat. The glass companies were very interested in creating a market for this new product, and they wanted a large building as a showpiece. In the case of the Hancock, the lead seal of the spacer bar was soldered to the glass on one side and to the coated glass on the other.

Several contributing factors led to the demise of the glass unit. The large panels with heat-absorbing coating underwent large thermal movements, which increased stress along the glass-to-tape bond. Eventually, the bond between the glass and the tape began to separate. However, the bond between the glass and the tape was so strong in some areas that the tape ripped microscopically small pieces of glass from the glass surface. The chips of glass along the edge were like stress risers that increased the stress levels in the glass. These areas became the weak point in resisting wind loads and ultimately were the source of the glass failure. When a storm produced high pressures on the glass, the amplified stress at the damaged edges caused the glass to break. The excellent bond between the solder and the reflective coating and between the coating and the glass precipitated the problem at Hancock. If the bond had not been so strong, the cracks in the solder might not have telegraphed through to the glass. It is ironic that the strong adhesion between these materials was the source of the glass failure at the Hancock. In the end, the entire façade was replaced with a new monolithic, fully tempered glass system. Today, the glass industry typically constructs insulating units with an aluminum spacer that is bonded to the glass with butyl sealant. This seal prevents air and moisture from entering the space between the glass and offers an improved connection between the spacer and the glass.

Wind Loads

The source of all of the enclosure failures mentioned above is to be found in the loads imposed on the system. Loads can include the dead load (typically the weight of the system itself), snow loads, seismic loads, maintenance loads, internal pressures from the mechanical system, and wind. The fracture of glass due to its brittle nature can be caused by various load factors, including impact, bending, point pressure, and thermal stress. Enclosure systems must be designed to resist all of these factors. The most common design load is wind. Two typical methods for determining wind loads used in the construction industry are analytical procedures and wind tunnel studies.

When buildings and structures have unusual geometric shapes or site locations, wind tunnel study is often used to determine the design loads. Wind tunnels can also be useful to establish a more accurate wind loading for large projects that warrant this level of preliminary investigation. The tests require thorough study of the existing natural wind variations near the site, a small model of the building and the surrounding areas appropriately sized for the study, and sophisticated instruments to measure the changes in pressure on the model.
Wind loads can be determined using a standardized formula, which considers a multitude of factors including the building's location, importance in the community, the size and shape of the building, surrounding conditions, and exposures. Forces are considered to be positive, inward acting, and negative, outward acting. Typically, the outward acting forces are the most severe on an enclosure. Enclosure elements are more susceptible to thrust loads than the building structure and hence must be designed for these higher forces. Predicting wind loads is critical to designing glass enclosures. Standard methods can be used to establish winds through an empirical formula that takes into account various factors including location, height of buildings, and surrounding conditions. Charts and graphs aid in illustrating that corners and eaves experience higher wind pressures than other areas of a façade. Basic wind speeds are taken from historical data in the area where the building is to be located. Sometimes this data is limited to areas of considerable distance from the site and can thus limit the validity of wind load estimates. How these wind speeds are translated into design pressures depends on local code requirements.

above The glass joists and posts are connected using resin adhesive in a mortise and tenor joint design.
Glass Museum, Kingswinford, England

page 123 A transparent box is achieved by using an all-glass structure.
Glass Museum, Kingswinford, England

Extreme weather conditions are often the cause of glazing failures. Typically, enclosures are designed for a 50-year storm wind load, occasionally for a 100-year storm. Additional safety factors are also incorporated into the design of the connections and materials.

Corners

Building enclosures must typically be designed for heavier loads at the corners and parapets. Negative pressure (wind pulling in an outward direction) is usually larger than positive wind loads (wind loads into the building). The negative wind loads traditionally govern, and are therefore used for testing.

An example of building failure due to high winds at the corner can be seen at the Ricola Building in Mulhouse-Brunstatt, France. Completed in 1993, the building was to be used simultaneously as a factory and for storage for this French herb candy manufacturer. The building form recalls a cardboard box lying on the floor with open flaps. The structure is 100 foot wide by 200 foot long (34.8 x 69.6 m). The 25 foot high (8.7 m) ceiling is supported by a steel frame and braced at its end by reinforced concrete walls. The cantilevered extending roofs on the two long sides both open up to the landscape and to the entrance and loading areas for forklifts and transport vehicles. They create shade and afford weather protection. The short sides of the factory building are each closed by the concrete wall. All roof water runs down over the concrete walls and trickles into a deep bed of gravel.

Both long walls are composed of translucent panels, which provide the work area with constant, pleasantly filtered daylight. The panels are a sandwich of honeycombed polycarbonate. Light filters through printed translucent polycarbonate panels, a common industrial building material. The design is progressive in the use of the building skin as a media screen. Using silk-screen, these panels display a repetitive plant motif. The image of an herbal leaf sprig is a company symbol.

Due to a windstorm in December 1999, the translucent panels were sucked free from their connection clips and displaced from the building. Some of the panels were damaged as a result of the failure. As predicted by a

fundamental wind study, the corner panels received the larger negative wind force and disconnected. Although these panels are standard building elements used in a slightly unconventional manner, they could not resist the forces of severe wind loads at the corners of the building.

Load Duration Factors

Annealed glass strength varies with the duration of stress which causes failure. This phenomenon decreases the effective load carrying capacity of glass when load duration is long. For example, if a snow load stays on an annealed skylight for two weeks, the load carrying capacity is reduced to approximately one-half of the one minute strength. Since heat-strengthened and tempered glass have residual surface compressions, which suppress the occurrence of tensile stresses, the reduction in strength is less pronounced; the remaining capacity is as follows: 60 % for annealed, 80 % for heat strengthened, and 95 % for tempered glass based on a two-week load duration. This reduction is reflected in the single light and insulated glass design procedures.

above The loss of panels on the building shows the need for stronger connections at the building corners. The most common mode of failure is an outward force at the corners. Ricola Building, Mulhouse-Brunstatt, France

above Translucent polycarbonate panels use silk screen images to decorate the façade. Ricola Building, Mulhouse-Brunstatt, France

right The loss of exterior enclosures poses a security risk for the building. Ricola Building, Mulhouse-Brunstatt, France

How can Structural Failures be Avoided?

Glass is a material that carries load in a simple linear fashion up to the threshold at which fracture occurs. The best glass enclosure designs are conceived so that there is a clear analyzable load path through the glass. Glass has evolved considerably since the construction of the Neue Nationalgalerie in Berlin. Recent projects have successfully used tempered glass as a load-carrying element from walls to roof structures. An example of a project that has used glass fins as structural members is the extension to the Glass Museum in Kingswinford, England. Completed in 1994, the glass roof of the extension is supported by triple laminated all-glass joists. The entrance is a transparent box, 11 m long and 5.7 m wide and 3.5 m high (36 foot x 18 foot 8 inch x 11 foot 6 inch). The 300 mm (12 inch) glass joists are designed without visible metal connectors. At the front of the entrance, the glass joists are connected to the glass vertical supports using resin adhesive. The connection executes mortise and tenor joints to form a rigid frame with resin adhesive. At the rear of the room, the glass joists are connected to an existing brick wall with steel joist hangers.

The key to the success of glass applying forces to glass is to create a clear path for the force, minimize concentrated loads, separate glass from metal with a forgiving material, keep the glass in compression, and provide accurate load predictions. There is no real substitution for glass. The projects described in this chapter that have had problems are the rarity. Glass manufacturers and experts are guiding the industry to safe and innovative solutions for our enclosures.

Lessons Learned

1
In its tempered form, glass can be stronger than steel, but it is still susceptible to localized over-stressing from acute loads. All materials in contact with glass should be soft enough and large enough to transfer forces evenly over this inherently brittle material.

2
Building enclosure elements must be analyzed for fatigue during the design process. The stresses that a material can tolerate under cyclic loading are much less than under static loading.

3
Poured and debridged polyurethane thermal barriers should be structurally tested to verify that the urethane formula has cured and that dry shrinkage will not occur after thermal cycling.

4
Laminated glass should not be considered to have the same strength as monolithic glass of the same thickness. Under sustained loads and high temperatures, laminated panes act in an independent manner and exhibit a loss in strength.

5
A classic problem in any manufacturing industry is introducing new materials which may have problems with a particular application. New technology often has growing pains. Even with all due care and a tremendous amount of professional responsibility, there is a chance that failure will occur. Consideration of a particular product's historical performance is an important factor when fabricating the same detail in a highly repetitive fashion over large areas.

6
Glass failures in enclosure systems are not always caused by a problem with the glass. Excess deflection of the supporting material or failure of the fasteners will lead to glass breakage.

7
Thermal cycling is an important element of mock-up testing. Exposing a system to thermal extremes can change their properties and ultimately affect their performance. If possible, the mock-up system should include thermal cycling followed by structural load and air/water infiltration testing. This procedure has been crucial to determining the effectiveness of thermal breaks of systems for many curtainwall projects. Structural load testing should be done before and after thermal testing to ensure that thermal cycling does not affect the structural performance of the wall.

8
There are inherent limitations and minor visual distortions in heat-treated glass that should be considered when specifying heat-treated glass. These characteristics include strain pattern and roller-wave distortion.

9
Cladding elements must be designed to consider the worst case loading for a building. Special design consideration for high winds at corners and eaves is typically required.

Redundancy

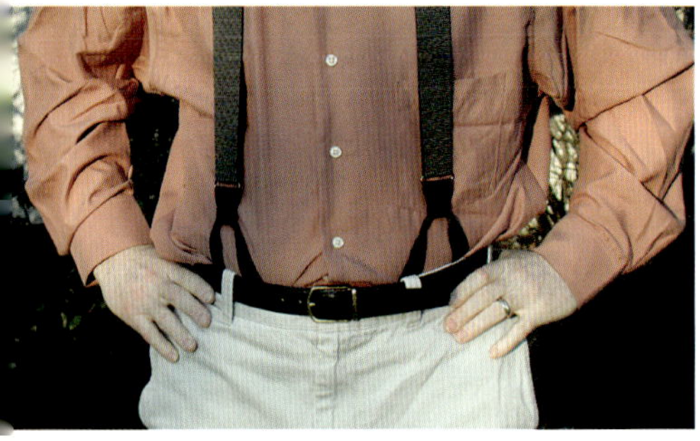

The incidence of building enclosure failure in relatively new buildings is often the result of a trend toward lightweight, inexpensive wall construction. In the past, building envelopes were made of massive, solid façades of brick and stone. Redundancy, sometimes characterized as a belt-and-suspenders philosophy, is an important consideration in enclosure design. Older structures incorporated built-in redundancy that provided a back-up system to help these walls resist premature degradation, even when they were not properly maintained. Many contemporary façades, however, rely on tenuous, marginal methods for resisting wind and preventing water from entering into the walls. All structural failures can be classified as failures due to insufficient redundancy. A common practice in building enclosure design is detailing with a certain level of redundancy in order to avoid failure. One method of reducing the risk of structural failure is to incorporate structural redundancy into the design. It is important to realize that this design strategy is independent of material safety factors. Material safety factors relate to one connection's performance and not the risk of failure of the system. Glass is used today with a greater sense of boldness. We are using it for our floors, for the treads on our stairs, and even for designing cantilevered canopies. This new confidence is not reckless but it does utilize a conservative level of redundancy in design.

above Redundant design is also termed a belt and suspenders approach.

right Mock-up testing uncovers an anchor failure which causes the entire unit to displace. Redundancy in design can mean that the loss of any one connection will not cause the displacement of the framing system from the building.

page 127 The interior glass skin folds inward as it reaches for the lobby ceiling. Shortly after completion, the glass wall was modified in order to achieve redundancy in the design.
Guggenheim Museum, Bilbao

Redundancy in design does not always apply to structural elements. Redundancy for a weather-tight enclosure can mean providing dual seals at all joinery locations. A dual seal design inherently sees to it that the loss of the outer seals will not introduce water into the building. Often, four-sided structural silicone sealant glazing is backed up with a mechanical fastener.

Redundancy in the design of structural support can be simplified with the understanding that the loss of any one connection will not cause complete displacement of an enclosure element. A simple example of this concept can be understood by examining the point-supported connections of a vertical glass panel. A glass panel with point-supported connections typically has four points of contact. Two of the supports may be dead load connections, while all four provide lateral support. If all four connections supported the dead load of the glass, the panel would be locked up and unable to accommodate any thermal expansion. If one of the two dead load fasteners should fail, the lateral connections will shift in order to support the displaced dead load. It is preferable for the loss of a con-

above The museum's shapely exterior envelope is similar in form to the interior atrium glazing skin.
Guggenheim Museum, Bilbao

center The replacement system provides redundant connections to the glazing support structure.
Guggenheim Museum, Bilbao

nection to change the orientation of the panel. This makes the failure apparent upon inspection. Because glass fractures into small pieces, redundancy for glass materials can mean providing multiple layers of glass or plastic sheets in order to provide structural support should any one layer break. Glass enclosures are fragile skins that are susceptible to failure. Built-in redundancies in these types of enclosures can prevent additional damage and injury. Glass breakage is not the problem. It is the given. The question is: what happens to the enclosure after an isolated break?

Completed in 1997, the Guggenheim Museum is a masterful architectural achievement both for the Basque population of Bilbao and the arts and architecture community throughout the world. Constructed primarily by Basque Contractors, the Museum is a remarkable design solution for the city, the people, and the art within. The building is famous for its unique three-dimensional free-form titanium skin. The inside of the building is equally enhanced with a shapely multi-storey wall made entirely of glass. The segmented glass skin separates interior balconies from a large multi-storey atrium space. The panes of glass are detailed like shingles, one connection tucked behind the other, as the wall curves up the atrium as a free shape. The glass used is tempered for safety. Two holes were drilled into the upper part of the glass to support the dead load of the material. The bottom of the glass was laterally supported at two points using structural silicone. The wall is not vertical but takes a concave shape up toward the atrium ceiling. Inadvertently, a sheet of glass in the enclosure broke and fell down like a waterfall breaking other pieces located in lower areas. Because of the wall's shape, a single failure caused a multitude of broken glass below. Broken panels fell to create more broken panels. The solution to the problem was to add a plastic film to the reverse side of the pieces. This way, if the

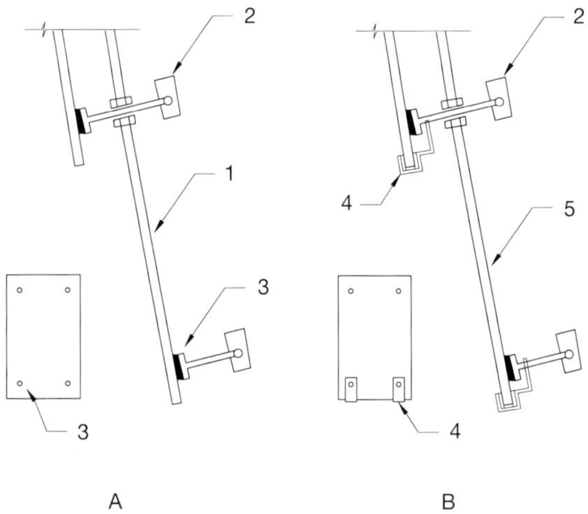

right The new façades include laminated glass that improves the acoustical performance and provides a level of redundancy to the glazing system.
Bracken House, London

center An isolated occurrence of glass breakage is aided by the laminated glass's ability to stay within the glazing frame. The blue tarp was added to prevent loosening glass fragments from falling to the street below.
Bracken House, London

page 128 below
A Original interior glass skin design
B Improved design with redundant features
1 Tempered glass
2 Support connection to steel frame
3 Silicone point connection to glass
4 Metal kerf added to provide additional support of glass
5 Adhesive film added to the glass to hold glass fragments together upon breakage
Guggenheim Museum, Bilbao

tempered glass breaks, the pieces will remain fixed to the film (similar to laminated glass). In addition, kerf-like supports were added to the lower connection. The kerfs would support the glass in case of failure. The replacement system adds great improvements to the redundancy of the system. Glass enclosures are fragile and will break. This chapter discusses the concept of redundancy in glazing design.

Laminated Glass and Added Films on Glass

Plastic films added to glass can be seen as a form of redundancy. If the glass is broken, the interlayer will resist the displacement of the glass from the frame. Because tempered (toughened) glass breaks into small pieces, it is not recommended to provide both lights of a laminated unit with tempered glass. One problem with double-tempered glass is the possibility of profound optical distortion. Thin tempered glass has some roller-wave distortion, and with two thin tempered lights laminated together, the result can be small lens-like distortions, which greatly reduces transparency of the glass. The second drawback to a double-tempered laminated light of glass is that if both lights are broken, the unit will possibly sag out of the frame. Laminated films are often added to glass enclosures in order to improve security by providing redundant support if the glass is broken.

London's Bracken House, former newspaper printworks and offices constructed in 1950, was converted into a bank in 1992. The "Techno-Gothic" solution included new façades for the center block of the building. The new façades were designed to reflect the clients' sophisticated level of service as well as their need to maintain security. Projecting bays of double-glazed panels in bronze frames are cantilevered from the main structure's reinforced concrete frame. The wall's load is transferred through delicate columns and hefty brackets, both made of cast gunmetal (an alloy of copper and tin), down to load-bearing stone piers at the building's base. Because the exterior light of glass is laminated, it will not displace if broken. This is important to a financial institution where security is a priority. If a light of glass should break, the film will temporarily hold the glass in place

above A broken tempered top light in a filigree glazing system
Greenhouse Parc Citroën, Paris

below Diagram of gravity load path of glazing system:
A Before glass breakage
B After glass breakage
Greenhouse Parc Citroën, Paris

page 131 The broken light does not affect the structural performance of adjacent glass panels.
Greenhouse Parc Citroën, Paris

until a replacement piece is installed. Laminated layers and plastic films can add a level of redundancy to enclosure systems.

Monolithic spandrel glass is typically required to meet fallout resistance testing. Because these units are not always visible from inside the building and can be difficult to see clearly from the street on tall buildings, a broken spandrel can go undetected. Scrim sheets of polyester film laminated to the spandrel glass with solvent-based adhesives are often used to prevent glass fallout after the glass has been damaged, thus providing a redundant mechanism in case of breakage. Scrim sheets are typically employed on the back of mirrors as a method of providing safety glazing without using tempered glass (which has some optical distortions).

Zipper Effect

There are several mullion-less class façades developed in recent years including truss-braced and cable-stayed. All of these highly transparent enclosure systems can suffer what is known as a zipper effect if the skin does not incorporate redundancy in the design. The zipper effect is a chain reaction of glass breakage, one failure contributing to the next, as catastrophic breakage rips through an enclosure.

In order to avoid catastrophic failure, each piece of glass must be independently supported. One light of glass cannot rely only on an adjacent piece because, due to the fragile nature of glass, that piece may not always be there. Since all glass is susceptible to breakage, one must provide solutions in case of failure. To the untrained eye, the large expanse of the filigree glazing systems may look like a house of cards, held together by sealant. It might appear that breakage of one glass light would cause failure of adjacent panels. In fact, these systems are carefully designed to redistribute load in the event that glass breakage should occur.

The Greenhouses of Parc Citroën, Paris, were constructed in 1992 on a site previously occupied by a Citroën car factory. The four small greenhouses are rectangular volumes measuring 8 m long x 6 m wide x 9 m high (23 x 17 x 26 foot). A variety of materials are used in the construction of the greenhouses. Stone, bronze, wood, and stainless steel cable tension elements fill the space created by the glass box. The enclosure exposes the plants within to light from all directions, including the roof. The top row of glass lights on the greenhouse is suspended from the peripheral beam by a single connection. The lights below appear to be relying on this connection for support. In fact, the system has been designed in anticipation of breakage of the upper lights. Breakage of a glass sheet for this

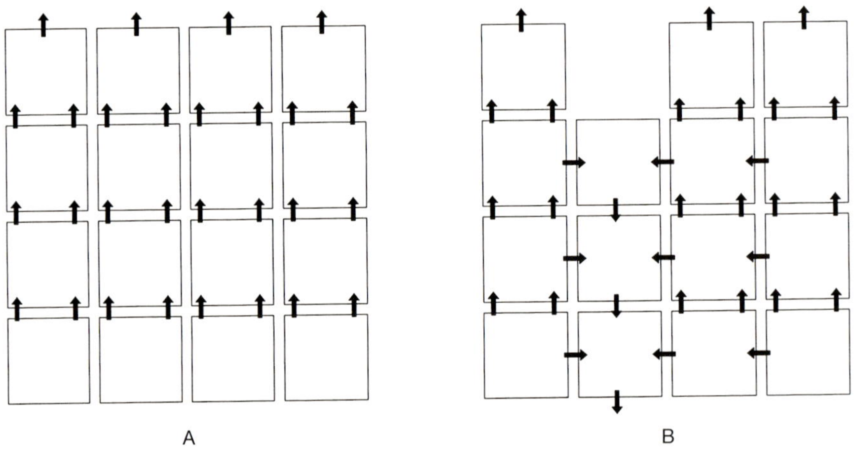

above and right The building skin is composed of unitized curtainwall elements that introduce terra-cotta tube elements to the façade.
Debis Tower, Potsdamer Platz, Berlin

type of system will cause an immediate change in the load path of the glass suspension system. Should the upper sheet in a row break, adjacent panels will support the remaining sheets. The loads are transferred through horizontal casting of the four-hole connection, and the shear stress is transferred through the silicone joints. Because a sudden load is far more likely to cause a fracture in glass, the wing-like springs at the top connection reduce the effects of such shocks. It is apparent by examining the breakage that the glass is tempered, but not laminated. This resulted in full displacement of the glass light upon failure. It should also be noted that the failure occurred at a corner panel, statistically prone to higher wind loads. The glass breakage on this project reinforces how the individual lights of the filigree façade are independently supported, and the loss of one light of glass will not cause the failure of another.

Cladding Elements

Redundancy can be seen in enclosure elements other than glass. The façade of Debis Tower at Potsdamer Platz in Berlin incorporates thin terra-cotta tubes in its design. Unlike the glazed terra-cotta commonly found on office buildings built around 1900, this unglazed material, developed in the 1980s, is made of carefully selected clay of higher density and greater strength than the traditional material. The rod-like sunshade elements, referred to as "baguettes", fasten into receptors cast into the vertical mullions. Galvanized metal fittings attach the cladding to the substructure. An aluminum reinforcement tube threads through the ceramic baguettes preventing their displacement if the terra-cotta breaks. This is a classic example of redundancy in curtainwall design; the inner aluminum tube provides a secondary line of support.

above right and below right The tower has a four-sided structurally glazed curtainwall. Bibliothèque Nationale, Paris

right Detail section through curtainwall Debis Tower, Berlin
1 Glass window
2 Terra-cotta "baguette"
3 Aluminum tube

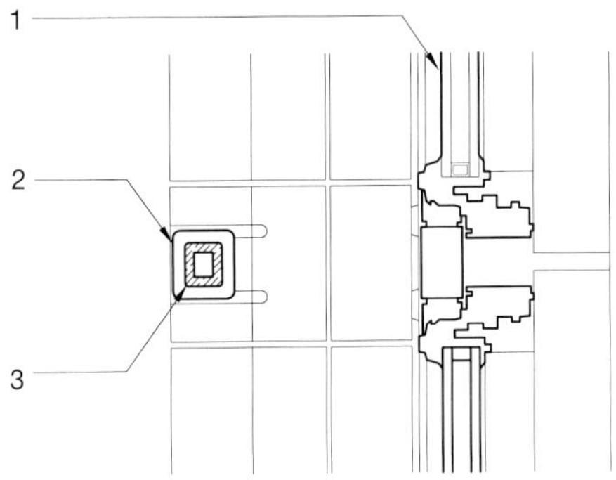

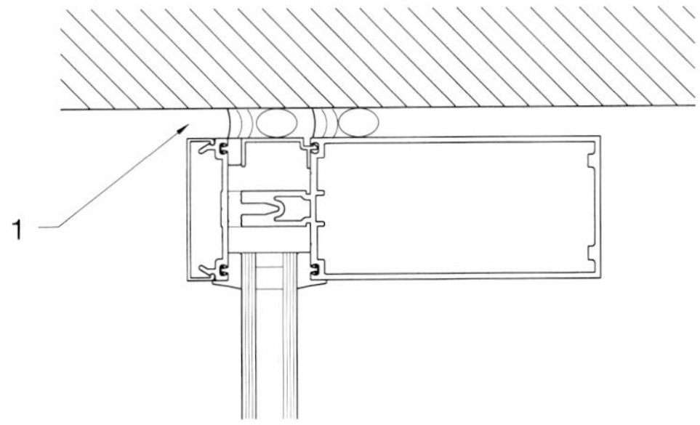

above Redundant joinery design: Dual seals (1) protect the interior primary seal from weather and the inside of the building from air and water infiltration.

below Redundancy for glass connection: Four-sided structural glazed curtainwall (1), corner attachments (2) provide a level of redundancy to the adhered glazing system. Bibliothèque Nationale, Paris

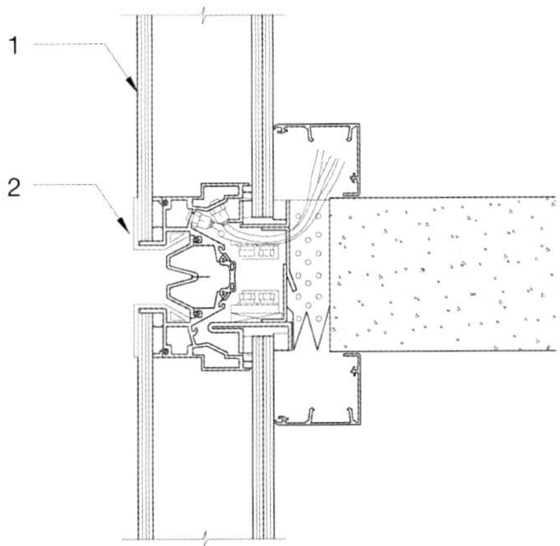

Dual Seals

Reliance on an exposed single barrier of sealant to prevent water penetration into an enclosure is the single most common source of contemporary building enclosure problems. Despite improved sealant technology and installation techniques, a single sealant barrier is only effective if it maintains its original seals. Because of all of the problems that can lead toward sealant failure, including weathering, poor workmanship during installation, incompatibility problems, improper mix design, preparation of substrates, or ultraviolet radiation, an enclosure design that relies on single seals lacks the level of redundancy required for enclosure system design. The concept of a dual-seal glass enclosure is another form of redundancy. This type of design provides a continuous outer and inner seal at all joint details. The outer seal can fail, and the system will still resist water and air infiltration. A dual-seal system allows the primary seal (the one on the inboard side of the system) to be sheltered from rain, sun, and birds. All of the these elements could adversely affect its performance. The outer seal can be seen as a sacrificial element.

Four-Sided Structural Silicone Sealant

The concept of adhering glass to metal on the outside of buildings has been a classic architectural detail since the 1970s. The detail relies on the silicone mixture to be exact and applied to properly prepared substrates. Complete sealant curing is required in order to obtain full adhesion. There are extensive quality control measures required for four-sided structural silicone sealant applications in order to ensure that the silicone mix, the surface prep and curing have all been executed properly. This quality surveillance typically occurs in a controlled environmental setting as most four-sided projects are glazed in a factory in a unitized or partially unitized system. Some people still believe that to rely on the sealant alone lacks the level of redundancy typically required for an enclosure. If something was missed and the silicone did not perform, the glass could fall off the building. Some designs install a mechanical connection to the enclosure frame as an added level of redundancy.

Completed in 1997, the four towers of the Bibliothèque Nationale in Paris are covered with four-sided structurally glazed curtainwall. The outer light of glass is 1.8 m wide x 3.6 m tall (6 x 12 foot). The dead load of the glass is supported by glazing blocks, which rest on the

right Glass treads are composed of a laminated composite material: annealed glass and acrylic panels
Rutland Gate Residence, London

below The multi-layer glass treads have built-in redundancy.
Rutland Gate Residence, London

page 135 The design includes a glass roof with triangular shaped glass and acrylic panels.
Yurakucho Canopy, Tokyo

aluminum frame. The weight of the glass could be supported by the silicone sealant, but most designs do not rely on the silicone to bear the dead load. The negative wind pressure on the glass is resisted by the sealant. This is what keeps the glass from being pulled off the building by suction winds. The corners of the windows have small clips that provide additional resistance to negative pressure. Clips like these can be added to the glass to provide a level of redundancy to the enclosure system.

Glass Walking Surfaces

Laminated glass can be used to ensure a level of redundancy in the structural capacity of load-bearing glass. Glass stair treads as well as glass floor systems are typically designed with redundant layers of glass to support

design loads even after a layer of glass has been broken. To accommodate the possibility of accidental damage to a layer of the laminated glass, the glass should contain at least one more layer than the design load requires. For example, if the mechanical analysis indicates that two leaves are sufficient, then three should be used. The extra, redundant leaf should be ignored for design purposes. All leaves of the laminated glass should be the same thickness.

Redundant glass designs have opened up exciting design possibilities for this material. The Rutland Gate project in London incorporated a spiral glass stair through the building. With glass treads and a light metal frame, the stair allows light to penetrate through the house from the top floor. Delicately detailed, the stair treads are composed of a composite section of 3/4 inch (19 mm) annealed glass and 5/8 inch (16 mm) thick acrylic sheets. The annealed glass is sand-blasted in a diamond pattern to give a slip-resistant finish to the top of the tread and is strong enough to take the design load for the stair. The acrylic sheet provides a redundant layer that has been designed to support the design loads should the glass fail. The glass and acrylic both contribute to the success of the system. The annealed glass provides the textured durable surface required for a floor surface, and the acrylic gives the flexibility and strength for a redundant design. The acrylic has a low modulus of elasticity, so it would deflect much

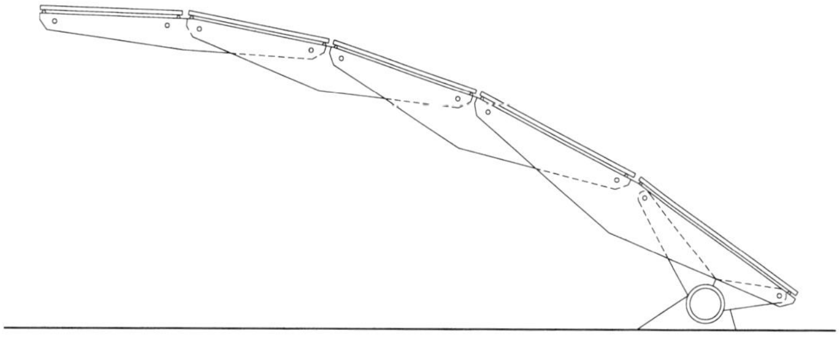

above The cantilevered canopy made of glass and acrylic panels has been designed to withstand the forces of an earthquake. Yurakucho Canopy, Tokyo

center The multiple glass blades have sufficient redundancy to cope with up to 75% of the elements breaking without the collapse of the canopy. Yurakucho Canopy, Tokyo

more than the glass if it were the sole support of the load. This composite stair section depicts a fundamental requirement of a well-designed redundant system. If the glass should break, the acrylic tread would support the weight of a person while deflecting like a safety net under the load. This deflection would alert the user that the top tread is broken and prompt a repair. If failed systems go undetected, the secondary systems often continue without a back-up.

Glass Structures

Inspired by the composite design of the Rutland stair, the Yurakucho Canopy in Tokyo uses cantilevered glass and acrylic panels to shelter the entrance to the underground system. Completed in 1996, the canopy is 10.6 m long, 4.8 m wide, and 4.8 m tall (35 foot 5 inch x 15 foot 9 inch x 15 foot 9 inch). The structure is supported by three parallel, cantilevered beams, composed of triangular shaped laminated blades (2 x 19 mm or 2 x 3/4 inch tempered glass) and 40 mm (1 1/2 inch) thick acrylic panels. The roof of the canopy is made up of 1.9 to 2.5 m long and 4.8 m wide (6 to 8 foot long x 15 foot 9 inch wide) laminated sheets (2 x 15 mm or 2 x 9/16 inch toughened glass) connected to the glass beams with counter-sunk stainless steel fasteners. The acrylic panels provide redundancy for a glass structure that would otherwise collapse in the case of massive glass failure. The acrylic panels were incorporated in case of an earthquake. The laminated multiple glass blades have the redundancy to cope with up to 75 % of the elements breaking without the collapse of the canopy. Similar to the Rutland stair, if the glass fins of the canopy should break, the acrylic fins can independently support the design loads of the canopy. The installation of a cantilevered glass canopy shows how far we have come with the use of glass in construction. Designers are growing more confident with glass, and the use of acrylic composite sheets has aided the process. In a similar vein, the University of Stuttgart is in the process of developing glass with integral carbon fibers to maintain the structural performance of the glass after breakage.

Glass floors are designed with multiple plies of laminated glass. The system is designed to support the loads even if one or more of the layers are broken.

How can Glass Enclosures be Designed for Redundancy?

Many specifications can require that a building enclosure system be designed and engineered with provisions for redundant load path. This means that failure of a single fastener, anchor, support, or framing member will not result in the destructive dislocation of any member or component, nor contribute to progressive collapse. One method of verifying the structural redundancy of an enclosure is through testing. A laboratory mock-up affords designers the opportunity to confirm that the loss of a structural connection will not cause displacement of the enclosure system. This test consists of intentionally removing an anchor support and applying the negative pressure design load. The test will confirm that the system can redistribute the load to other anchors without displacement of the enclosure.

The belt-and-suspenders design philosophy reminds us to hope for the best, but plan for the worst. Enclosure designs must anticipate that glass will break, and that seals will fail. The question is: what happens next? Will the occurrence be isolated, or will it start a catastrophic failure? Designers have found a new confidence in using glass for building elements. Glass is being employed for floor systems, stairs, and other load-carrying elements. These daring new designs do not move in this direction recklessly, but incorporate redundancy into their systems.

Lessons Learned

1
For many enclosures, the question is not whether or not the glass will break. The issue is what happens when it breaks. Systems cannot eliminate the possibly of glass failure, but must be designed to accommodate that failure when it occurs. Failure of any one element (for example, shattering of a glass window) should not cause a "zipper effect". Each light of glass must be independently supported.

2
Overhead glazing should be laminated and if tempered (toughened) glass is used because of a strength requirement, at least one of the laminated lights of glass should be heat-strengthened. This will prevent displacement of the glass in case of glass fracture.

3
Exterior sealant failures should be anticipated in enclosure systems. A dual seal design will prevent air and water infiltration into the building.

4
Plastic films are often added to enclosure systems as an added level of security. The film's ability to temporarily hold the broken glass together provides redundancy to the system.

5
The use of acrylic sheets in conjunction with glass creates a composite structural system that is durable and safe.

Impact

Glass is a fragile material susceptible to impact damage. Durability of a glass enclosure depends not only on the forces applied to it, but also on how it is put together and protected. Building enclosures are vulnerable to a variety of unwanted attackers. Because of its ability to bond glass fragments together and resist impact, laminated glass is ideal for providing protection against thrown stones, striking axes, and speeding bullets. Whether the object is slow-moving and large or fast and small, glass systems can be designed to protect the inhabitants within. There are various levels of resistance which glass can be expected to deter. Depending on the forces it is trying to resist, a barrier of laminated glass can be composed of various layers of thermally strengthened glass, wire, protective films, plastic sheets, and cast-in-place resins to achieve the level of protection desired. Lamination of glass will not prevent breakage; however, it will prevent further damage to other units by keeping the glass in place. Car windshields are laminated to reduce the risk of injury to passengers if a flying object hits them. Many architects are designing enclosure systems with laminated glass to minimize the damage caused by impact loads. In addition to greater durability, laminated glass provides a thermal and acoustical benefit to the glass skin.

Hurricane season churns out tropical storms around the world. Many coastlines are routinely battered by these storms, and although nothing can be done to prevent them from striking, there have been studies to develop economic methods of minimizing the damage to buildings. The primary source of damage to building skins is not the high winds but flying objects in a storm. It is not uncommon for a domino effect to occur, where flying debris breaks the glass, which becomes flying debris

above Glass spandrel damaged by impact from window washing stage.

right Glass can be designed to stop bullets.

page 139 Glass breakage due to the collapse of the World Trade Center, New York

and breaks more glass. Research is aiming to provide a sacrificial ply to enclosure systems to prevent the chain reaction of broken glass. The concept is fairly simple. Two plies of glass are laminated to a thin clear plastic sheet to produce a laminated glass window unit. During a tornado, typhoon, or hurricane, the outer exterior-facing glass ply is sacrificed to the impact of flying debris. The inner glass ply and water seals around the glass perimeter are maintained, protecting the building and its occupants. Ironically, the use of laminated glass to protect building enclosure systems from windborne debris originated not from buildings in a hurricane region, but from a history of glass failure in the city of Chicago.

Chicago's Sears Tower was heralded as the world's tallest building after its completion in 1973. Along with its impressive title came a history of glass breakage on the curtainwall façade. Shortly after completion, winds of 75 mph (120 kmh) shattered 16 windows on its upper floors. As the panes fell to the ground, they damaged 40 more windows below. In February of 1988, 90 panes were broken during a windstorm with gusts of 56 mph (90 kmh). In a storm the following April, 97 more windows were lost as winds whipped through the city at 75 miles per hour. Plywood sheets were used to temporarily replace the 5 x 8 foot (1.5 x 2.4 m) windows. The building enclosure had been designed and tested to resist wind velocities of 150 mph (240 kmh). Many people understood that the winds alone were not the only factor in the breakage. The two occurrences in the same year seriously challenged claims that the breakage was due to freakish acts of nature. Although the broken units represented a small percentage of the 16,000 windows in the tower, a solution to reduce the risk of another incident was required. Windborne debris was attributed as the main cause of the glass breakage. Subsequent damage was believed to be the result of a "cascade effect". Minute surface erosion and scratches from falling glass can reduce glass panes' strength below. The cascade of falling glass can create a pattern of future breakage.

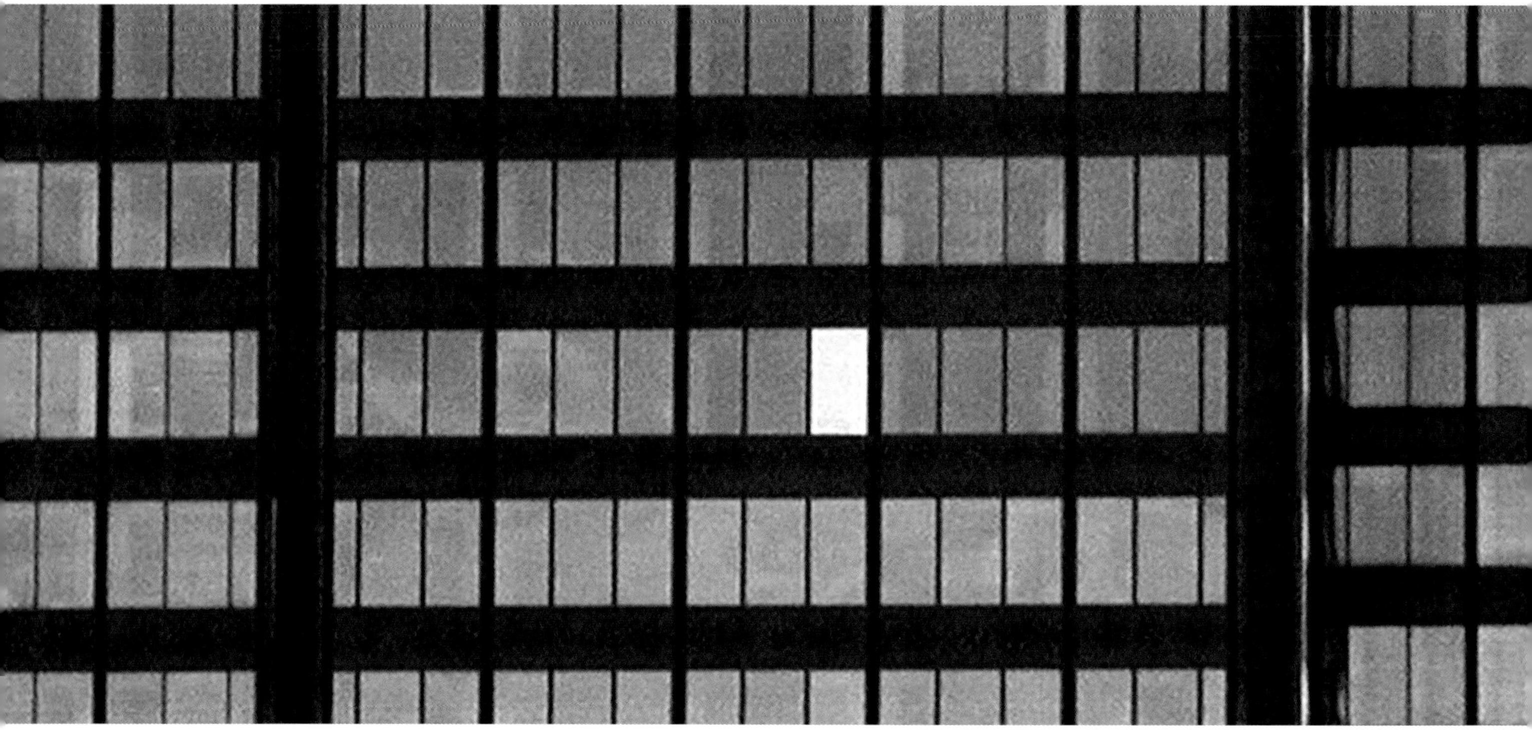

Temporary replacement lights do not always match the tower's original glass color. Sears Tower, Chicago

Completed in 1973, the Sears Tower has had a history of glass breakage that has been attributed to flying debris.
Sears Tower, Chicago

The solution to the problem of broken glass creating more broken glass was achieved by a sophisticated replacement system started in 2001. The glazing was systematically removed from the building and replaced with a custom frame that encapsulates a laminated light of glass with a reflective coating. The reflective coating is on the inboard side of the exterior light of heat-strengthened glass. Because glass conducts heat relatively efficiently, increased resistance can be achieved by introducing layers of non-conductive material between sheets of glass. This explains how the introduction of PVB (polyvinylbuyral) interlayer improves the thermal performance of the monolithic enclosure. The reflective coating on the inboard side of the outer light of glass reduces the solar heat gain of the system. The replacement system is virtually undetectable from the street but offers many added benefits to the original building skin. In addition to holding the glass in place if broken, the laminated glass with a reflective coating provides an improved thermal performance to the system. Tall buildings must be designed to withstand the high pressures of wind, but building designers may also need to begin accounting for flying debris.

Hurricanes

Typhoons and hurricanes have wreaked havoc on human enclosures long before the use of glass on building façades. In order to comprehend the forces involved with these extreme environment conditions, one must study the wind speed associated with hurricanes. The Saffir-Simpson Hurricane Scale rates hurricanes by intensity from 1-5. The scale is used to give a comparable estimate as to the potential damage expected from a hurricane landfall. The wind speed is the determining factor in the scale. All wind speeds are based on a one-minute average.

Category 1
Winds 74-95 mph (119-153 kmh)
No real damage to building structures

Category 2
Winds 96-110 mph (154-177 kmh)
Some roofing material, door and window damage of buildings

Category 3
Winds 111-130 mph (178-209 kmh)
Some damage to small residences and utility buildings with a minor amount of curtainwall failures

Category 4
Winds 131-155 mph (210-249 kmh)
More extensive curtainwall failures with some complete roof structure failures and small residence failures

Category 5
Winds greater than 155 mph (249 kmh)
Some complete building failures, with small utility buildings blown over or away

Typhoon-Resistant Glass

The use of laminated glass in high-rise buildings throughout the world has become popular for a variety of reasons including noise attenuation, protection against glass fallout, solar control, and typhoon resistance. The Cheung Kong Center in Hong Kong uses 40,000 square meter of laminated glass around its 60-storey towering enclosure. The building is situated in one of Hong Kong's busiest streets, Queen's Road Central. The design responds to feng shui principles to determine the building's orientation, square plan configuration, and use of highly reflective cladding materials. An acoustically improved enclosure was a priority for this noisy urban environment. The glass units are composed of a 10 mm (3/8 inch) clear, heat-strengthened glass pane, followed by a reflective coating, a 1.52 mm (1/16 inch) PVB interlayer, and a 6 mm (1/4 inch) heat-strengthened inner pane. The total unit thickness is 17.52 mm (11/16 inch). It is interesting to note that tempered glass is not used. Following a major typhoon in 1999, fewer than ten pieces of glass had to be replaced. These repairs were minor as compared to buildings around the tower which suffered major damage. Because the glass of the Cheung Kong Center was contained within the frame, falling glass during the typhoon did not damage additional glass units below. Standard annealed or tempered glass would have fallen out of the frame, potentially causing damage to other windows or people on the street. Typhoons are not the only danger to glass buildings. Many forms of aggression must be considered in designing a building enclosure.

right Aftermath of a Florida hurricane

below left "Star Burst" Localized thinning of the PVB Interlayer of a laminated light of glass during the manufacturing process can lead to an aesthetic defect.

below right "Fingers" A common manufacturing flaw in laminated glass is small air bubbles that can be blown between the lights of glass during the manufacturing process.

page 143 Laminated glass is used for the exterior skin of this tower to minimize the damage caused by typhoons. Cheung Kong Center, Hong Kong

CITIBANK

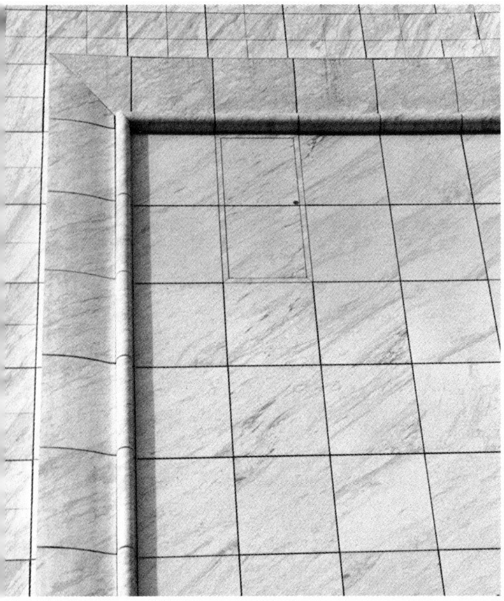

Anti-Vandal Glazing

Don't throw stones if you live in a glass house. Vandalism is a serious threat for glass structures. Anti-vandal glass has a range of different protection levels in anticipation of a building's needs. Classified as type "A", anti-vandal glass is intended to prevent the penetration of stones. This is tested by dropping a 9.1 lb (4.11 kg) steel ball three times in succession onto the glass from various heights [A1: 11 foot 6 inch (3.5 m), A2: 21 foot 4 inch (6.5 m), A3: 31 foot 2 inch (9.5 m)], representing differing levels of protection. For instance, an A3 glass might be selected for the ground level, and an A1 glass for upper-level windows. Laminated glass, including several layers of film or suitable cast-in-place resin, can achieve this level of protection.

above left Stone façade is actually a glass curtainwall with a laminated outer layer of stone. ENSAD, Paris

above right Vandals prove the building is glass rather than stone with their own stones. ENSAD, Paris

below Translucent stone façade allows light into the building. ENSAD, Paris

An example of a project that could have used a type A1 classification for the glass enclosure is ENSAD – Ecole Nationale Supérieure des Arts Décoratifs, Paris. Completed in 1997, the translucent marble façade of this French school of design has been the target of vandalism. A fine layer of marble is laminated to the glass of a traditional curtainwall system, making a translucent stone façade. The fabrication of these glass units requires the exterior stone layer to be polished down to a very thin layer in order to improve light transmittance. The panels are susceptible to vandals that may want to prove the building is glass and not stone. The glass can easily be broken but is very difficult to replace.

Architectural glass elements that are separated from occupied buildings are particularly susceptible to vandalism. The glass floor panels of the Campo Volantin Footbridge completed in 1997 have also been the target of vandals. This inclined parabolic arch structure has a span of 71 m (233 foot). Serving to link a run-down commercial area to the city center of Bilbao, the footbridge is one aspect of a large campaign of urban renewal. This beautiful structure gives the visitor the sense of frozen movement. The effect is heightened by the lightness and delicate detail of the steel elements. Its spectacular illumination by night emphasizes the importance of the bridge form. A translucent glass walking surface allows for concealed lighting to illuminate the structure above. An unlikely enemy has

The bridge form resembles an inclined parabolic arch. The floor of the bridge is composed of glass panels. Campo Volantin Footbridge, Bilbao

below A shattered panel is near what appears to be a mock-up for a more slip-resistant replacement panel. Campo Volantin Footbridge, Bilbao

right Many glass panels have been shattered. Some have been temporarily replaced with wood infill panels. Campo Volantin Footbridge, Bilbao

undermined this incredible architectural sculpture. Vandals, perhaps unhappy with the bridge's slick glass walking surface, have taken it upon themselves to remedy this perceived problem by fracturing the panels. Because the small panes are laminated, they pose no immediate threat after being broken and can be walked on without fear of them falling out. This act of vandalism encourages the replacement of the existing panels with a tactile surface. In the meantime, the fractured tempered laminated glass provides an interim solution as a slip-resistant surface for pedestrians.

Anti-Intruder Glazing

Anti-intruder glazing is intended to prevent an opening larger than 400 x 400 mm (15 3/4 x 15 3/4 inch) pro being smashed in the glass with an axe. Classified as type "B", anti-intruder glass is intended to stop the entry of an individual into a building. Testing for this classification simulates a forced entry with a long-handled axe fitted to a machine. With the combination of glass and film or glass and poured-in-place resin, laminated glass provides the elasticity

to absorb the impact of the attacker. In order to be effective against intruders, a glass enclosure system must be designed in such a way that it cannot be pried out of the frame. For this reason, external glazing pressure bars should not be used.

Laminated glass at the ground floor of the Paris library shows the appropriate instance of type "B" anti-intruder glazing. Bibliothèque Nationale, Paris

right Traditional glass and glass block will not stop the penetration of a bullet.

Bullet-Resistant Glazing

Bullet-resistant glazing must be combined with a frame of the same level of impact resistance in order to be effective. Classified as type "C", bullet-resistant glazing is intended to stop a very small mass (a bullet) acting with a very high velocity (fired from a rifle). In order to be effective, the glass's mass is more important than its elasticity. As a result, bullet-resistant glazing contains a high proportion of glass. The classification varies depending on the type of weapon and ammunition used. The goal is for the bullet not to penetrate the glass and for no splinters of glass to be flung off from the rear surface of the glass. Bullet-resistant glass is composed of a thermally or chemically strengthened glass. Protective films or plastic layers are used to prevent the spalling of glass fragments. The design of the glass allows for the impact energy of the bullet to be transmitted through the material and released at the other end. Additional layers are added in order to achieve the no-spall designation.

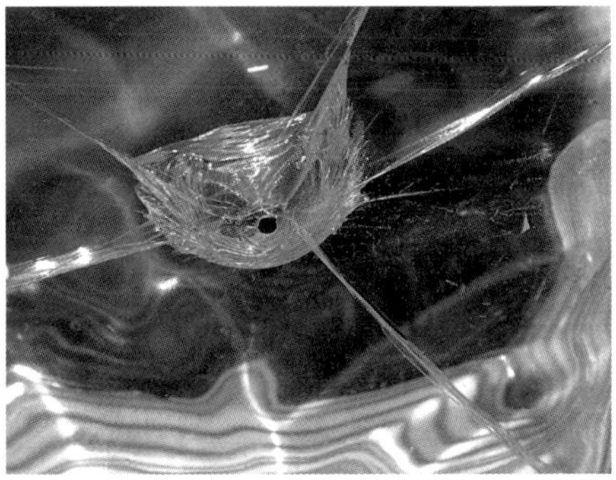

Explosion-Resistant Glazing

Explosion-resistant glazing is intended to protect against attacks from outside using explosive devices. Classified as type "D", explosion-resistant glazing can be composed of laminated safety glass. The design assumes spherical pressure waves acting perpendicular to the glass in the order of 4 psi or 576 psf (28 kPa). The elastic property of laminated safety glass makes the product suitable for this type of application. Like bullet-resistant glazing, the frames that hold the glass must meet the same level of impact specification as the glass.

On September 11, 2001, two commercial jet liners impacted the World Trade Center Twin Towers and devastated New York City. Clearly, no enclosure system could be designed to

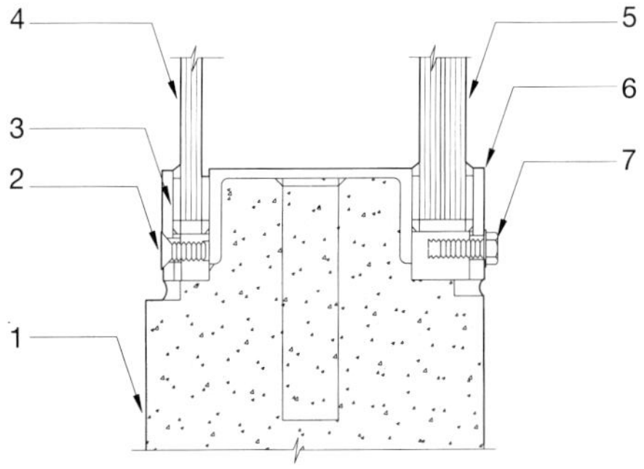

above Blast-resistant glazing detail:
1 Concrete panel with embedded steel channel
2 Screw fastener
3 Gasket with silicone sealant above
4 Sacrificial light
5 Blast-resistant glass
6 Steel plate
7 Bolt

right Unprotected glass edges provide crisp corner details but increase the risk of breakage prior to installation.

below Window washers' equipment should be made of hard plastics. Metal surfaces on tools can inadvertently damage glass.

withstand the forces involved in this attack. Many of the adjacent buildings, which did not suffer from direct contact, were damaged by the impact of the falling towers. There were many indirect casualties from falling glass of adjacent buildings. Fleeing pedestrians as well as incoming rescue workers were in constant danger of falling glass from partially damaged windows around the disaster site.

Casualties in a similar attack on the Pentagon in Washington, D.C. on September 11 were reduced because of many new improvements to the building including explosion-resistant walls and glass enclosure systems. Laminated glass is an improvement to enclosure protection against explosions; however, by itself, may not be enough. Shattered laminated glass is very flexible because the interlayer has no inherent bending strength. The fragmented pieces of glass provide little support and under blast pressures the laminated panel can fold up and be pushed out of its frame. Specially designed glazing frames can reduce the risk of glass fallout. With deep rebates and clamping devices, laminated glass can be retained in the frame after an explosion.

Many of the conditions described in this chapter are extreme and rare forces that a building skin can be exposed to. There are, however, many potential dangers lurking around an enclosure from the first days of its inception.

Impact of Construction

During the construction process, building enclosure units with glass are subjected to potential damage from many directions. The most susceptible areas of a glass pane are its edges. In order to protect against damage during installation, glass enclosures can be encapsulated in metal frames; however, some designs desire the crisp edge of glass to be open to view. Unprotected edges can easily receive the force of an acute load during shipping and installation of the units. Exposed glass edges are susceptible to breakage before and after installation.

above This blast-resistant enclosure design uses annealed glass in lieu of stronger tempered glass because it is more flexible and hence able to absorb the impact of an explosion better.
Lloyd D. George Federal Building and United States Courthouse, Las Vegas

page 149 Curtainwalls with exposed glass edges are very susceptible to damage during installation.

Welding splatter, sandblasting, wind-blown dirt or sand, or other scratch-producing objects may damage glass during the construction process. Concrete spillage from above containing course aggregate can damage installed glass. Because this type of damage can reduce glass strength and increase the likelihood of glass breakage, lights with these types of defects must be replaced. Welding near installed glazing should be avoided, and if required in isolated instances, great care and protection should be provided to ensure that weld splatter does not come in contact with the glazing.

Building maintenance can be hazardous to a building enclosure. Window washers and their equipment pose a particular threat to installed systems with exposed edges. Hard metal surfaces from tools and equipment can damage building skins after installation. Protection from hard surfaces contacting glass is critical to preventing impact damage.

How can Buildings be Protected from Impact?

Lessons Learned

It is important to note that in all cases of the protective measures described, replacement of the glass after impact will be required. The glass will break. The goals of the systems are safety and security of people and property near the enclosure. Some of the design parameters described are extreme; however, they do have applications. In the wake of the 1995 Oklahoma City bombing, the design of a new courthouse in Las Vegas considered blast-resistant curtainwall for its enclosure. The goal was to heighten the level of protection for the building inhabitants. The blast-resistant glazing is a 1 inch (24 mm) thick insulating unit composed of an annealed exterior light, a 1/2 inch (12 mm) air space, and a laminated interior light held in place by an aluminum frame. The inboard light is composed of a polyvinyl butral layer between two sheets of 1/8 inch (3 mm) thick annealed glass.Two full-scale mock-up specimens were constructed for bomb-testing during the design process. Completed in 2000, the Lloyd D. George Federal Building and United States Courthouse, Las Vegas, was the first building in the United States to provide a curtainwall enclosure designed to resist bomb blasts. This state-of-the-art blast-resistant enclosure design uses annealed glass in lieu of the stronger tempered glass because it has more flexible properties, which absorb the impact of the explosion.

1
Glass is a fragile material that is susceptible to breakage from a variety of forces. Special laminated products can reduce the risk of damage to property and injury of people after the glass is broken.

2
Glass breakage on a building with displacement fragments can weaken glass below with minor scratches and abrasions. These defects can lead to future failures followed by additional scratches and abrasions. One method of reducing the endless cycle of breakage is the use of laminated glass, which does not displace upon breakage.

3
Glass enclosures can be designed to resist a variety of forces from rocks to bullets to small bombs. Building owners must weigh the risk of threats as compared to the cost associated with these protective products in order to select the envelope design most appropriate for their building.

4
Damage from hurricanes may come more from flying debris than from high winds. The use of laminated glass in regions with a history of typhoons can reduce the amount of glass debris, which contributes to damage and injury.

5
Enclosure details are susceptible to a variety of forces during construction. Protection of glass edges prior to installation reduces the risk of damage to this fragile material. Glass details that have exposed glass edges should be understood to be a risk for glass damage during and after installation.

6
Care should be taken to avoid contact with hard metal objects during the enclosure maintenance process. Tools should be covered in plastic and metal equipment should be wrapped with soft padding in areas susceptible to contact with glass.

Acknowledgments

A book that describes the lessons learned from glass enclosure problems could not have been completed without the cooperation and shared knowledge of many design professionals and people from the glass industry. I would like to extend my gratitude to those architects, fabricators and individual experts that contributed information to my research. This book was intended to educate people about glass breakage, with the intent to reduce the risk of problems in the future. Without the help of people in the industry it would not have been possible. Special thanks go to Ria Stein for faithfully advising me on the content, A.A. Sakhnovsky for graciously reviewing the text prior to publication, Gabrielle Pfaff for elegantly laying out the chapters and Caroline Behlen for patiently translating the text into German. My final thanks go to the University of Illinois at Urbana Champaign School of Architecture, College of Fine and Applied Arts, who kindly sponsored this publication, and the Plym Traveling Fellowship made possible through the generosity of Francis J. Plym and the endowment he established at the University of Illinois at Urbana Champaign School of Architecture in 1911.

About the Author

Patrick Loughran was born in 1964 and grew up in Oak Park, Illinois. He studied Civil Engineering at the University of Notre Dame in South Bend, Indiana, where he received his Bachelor of Science in 1986. He then obtained a Master of Architecture from the University of Illinois at Urbana Champaign, Illinois in 1990. He proceeded to work for several architectural practices in Chicago and has been employed at Lohan Caprile Goettsch since 1994 where he is responsible for the design and detailing of curtainwalls, skylights and canopy structures. In 1999, he traveled to Europe on a Francis J. Plym Traveling Fellowship and was awarded the Young Architect Award from the American Institute of Architects one year later. "Falling Glass", which is his first book, summarizes several years of research on building enclosure problems.

Selected Bibliography

Foreword

Slessor, Catherine; *Sustainable Architecture and High Technology*: Eco-Tech, Thames and Hudson Ltd., London, 1997.

Nickel Sulphide

Aldersey-Williams, Hugh; "End of the Line", *Architectural Record,* June 1994, p 90–97.

American National Standard for Safety Glazing Materials Used in Buildings – Safety Performance ANSI; *Specifications and Methods of Tests,* American National Standards Institute, ANSI Z97.1-1984 (R1994).

Barry, J.C.; "An Electron Microscopic Study of Nickel Sulfide Inclusions in Toughened Glass", *Journal of Materials Science,* Kluwer Academic Publishers, Norwell MA, Volume 36 No. 15, August 1, 2001, p 3721–3730.

Compagno, Andrea; Intelligent Glass Façades: *Material Practice Design;* Birkhäuser, Basel/Berlin/Boston, 1999.

Greenburg, Charles B. and Edward J. Stofka, "The Two µNi1-x S Transition in Glass", *PPG Technology Journal,* Volume 1, Number 1, PPG Industries, 1995, p 43-51.

Krewinkel, Heinz W.; *Glass Buildings: Material, Structure and Detail;* Birkhäuser, Basel/Berlin/ Boston, 1998.

Nashed, Fred; *Time-Saver Details for Exterior Wall Design,* McGraw-Hill, New York, 1995.

Schwartz, Thomas; "How to Avoid Glass Fracture", *Glass Digest,* March 15, 1984, p 58-61.

Thermal Stress

Amstock, Joseph S.; Handbook of Glass Construction, McGraw-Hill, New York, 1997.

Blaser, Werner; Mies van der Rohe: *The Art of Structure;* Architectural Publishers, Artemis Verlag und Verlag für Architektur, 1965.

Technical Service Report No. 130, PPG Industries, July 13, 1990. Wacek, Dan; "Glass Breakage, VI", *Viracon,* 1985.

Reese, Nancy I.Z. and Phil Geib; "Danger From Above", *Chicago Tribune,* July 16, 2000, p 14-15.

Schwartz, Thomas; "Analyzing the Cause of Fracture, After the Fact", Glass Digest, April 15, 1984, p 66–74.

Corrosion

"Dawn of the Bronze Age", *Building Design Envelope,* September 1999.

Freedman, Sidney and N.R. Greening; "Glass Stains, Causes and Remedies", *Modern Concrete,* April 1979, p 36-72.

Hoke, John Rat; *Architectural Graphic Standards,* Ninth Edition, Ramsey/Sleeper, John Wiley & Sons, Inc, New York, 1994.

Holgate, Alan; *The Art of Structural Engineering: The Work of Jörg Schlaich and his Team,* Axel Menges, Stuttgart/London, 1997.

Russell, James A; "Project Diary, Jewish Museum, Berlin", *Architecture Record,* January 1999, p 76–91.

Nashed, Fred; *Time-Saver Details for Exterior Wall Design,* McGraw-Hill, New York, 1995.

"Care and Storage of Glass", Technical Service Memo, PPG Industries, November 1, 1970.

Rush, Richard; "A Stain in the Pane", *Progressive Architecture,* August 1979, p 94–99.

Taylor, George; "Water, Water", *Glass Magazine,* April 1997, p 22–26.

"Windows fail steel-frame weathering test", *Engineering News Record,* December 10, 1970.

Zahner, L. William; *Architectural Metals,* John Wiley & Sons, Inc. New York, 1995.

Incompatibility

Hoke, John Rat; *Architectural Graphic Standards,* Ninth Edition, Ramsey/Sleeper, John Wiley & Sons, Inc, New York, 1994.

Rice, Peter and Hugh Dutton; *Structural Glass,* E & FN Spon, London, 1997.

Russell, James; "There and not There" (Article on the Grand Louvre Pyramid), *Architectural Record,* May 1988, p 144–148.

Leakage

AAMA 501-83 Methods of Test for Metal Curtainwalls, Architectural Aluminum Manufacturers Association, 1983.

Dow Corning, *Weatherproofing Sealant Guide,* Dow Corning Corporation, October 1994.

LaTona, Raymond W. and Thomas Schwartz; "Cladding Systems Against the Wall", *Architecture,* May 1990, p 129–131.

LeVoguer, Brian; "Details and Execution: The Road to Quality Air Barriers", *Construction Canada,* July 1996, p 10–12.

Kerr, Dale D., "Controlling Rain and Wind", *Architecture,* October 1994, p 117–119.

Kudder, Robert J. and Jeffrey L. Erdly, editors; "Water Leakage Through Building Façades", ASTM, 1998.

Ting, Dr. Raymond; "Evolution of Curtain Wall Design Against Water Infiltration", January 1997, p 34–36.

Sturdevant, John R.; "What Makes a Good Curtainwall?", *Progressive Architecture,* February 1994, p 70–77.

Energy

Bahr, Christian; *The New Bundestag in the Old Reichstag, Jaron,* Berlin, 1999.

Compagno, Andrea; *Intelligent Glass Façades: Material Practice Design;* Birkhäuser, Basel/Berlin/Boston, 1999.

Edwards, Brian; *Green Buildings Pay,* E & FN Spon, London, 1998.

Josef Gartner GmbH; *Double Skin Façades,* Gundelfingen, 1998.

Josef Gartner GmbH; *Glass Louver Sun-shading,* Gartner System, Gundelfingen, 1998.
Kieren, Martin; *New Architecture Berlin 1990–2000,* Jovis, Berlin, 1998.
Martin, Hervé; *Guide to Modern Architecture in Paris,* 1996.
Vonier, Thomas; "Culture Crash", *Progressive Architecture,* September 1995, p 62–67.
Humm, Othmar and Peter Toggweiler; *Photovoltaics in Architecture,* Birkhäuser, Basel/Boston/Berlin, 1993.

Structure

AAMA TIR-A9-1991, Metal Curtain Wall Fasteners, Architectural Aluminum Manufacturers Association, 1983.
Berger, Robert E.; "Dry Shrinkage in Polyurethane Thermal Barriers", *U.S. Glass Metal & Glazing,* January 1992, p 42–45.
Berger, Robert E.;"New Applications Raise Questions About Structural Polyurethane Thermal Barriers", *Glass Digest,* February 15, 1989, p 94–96.
Button, David and Brian Pye; *Glass in Building,* Pilkington Glass Limited, Butterworth Heinemann, Oxford, 1993.
Campbell, Robert; "Learning from the Hancock", *Architecture,* March 1988, p 68–75.
Cohn, David; "Ricola Europe, Mulhouse, France", *Architectural Record,* May 1995, p 88–89.
Davies, Colin; "Law Courts Addition, Palais de Justice", *Architecture,* January 1999, p 63-73.
Fronek, Steve; "Curtainwall Anchorage", *Glass Magazine,* June 20, 1994, p 20–26.
"A guide to the Structural Performance of Laminated Architectural Glass with Saflex Plastic Interlayer", Monsanto Company, St. Louis MO, 1988.
Permasteelisa; *Habitat, Technology and Architectural Envelopes,* 1999.
Rice, Peter and Hugh Dutton; *Structural Glass,* E & FN Spon, London, 1997.
Ryan, Raymund; "Law Courts, Bordeaux France", *Architectural Review;* July 1999, p 48–55.
Sutro, Dirk; "Into the Tunnel"; *Civil Engineering Magazine,* June 2000.
Schittich, Staib, Balkow, Schuler, Sobek; *Glass Construction Manual,* Birkhäuser, Basel/Berlin/Boston, 1999.
Schwartz, Thomas A.; "The John Hancock Tower Glass Failure – Debunking the Myths", *Proceedings of May 2000 ASCE Structures Congress,* Philadelphia, 2000.
Schwartz, Thomas A.; "When Bad Things Happen to Good Buildings", *Architecture Boston,* Winter 2000.

Redundancy

Compagno, Andrea; *Intelligent Glass Façades: Material Practice Design,* Birkhäuser, Basel/Berlin/Boston, 1999.
Rice, Peter and Hugh Dutton; *Structural Glass,* E & FN Spon, London, 1997.
Russell, James; "Criticism, Debis Tower", *Architectural Record,* October 1998, p 124–135.
Van Bruggen, Coosje; Frank O. Gehry, *Guggenheim Museum, Bilbao,* The Solomon R. Guggenheim Foundation, New York, 1999.

Impact

Button, David and Brian Pye; *Glass in Building,* Pilkington Glass Limited, Butterworth Heinemann, Oxford, 1993.
Daniels, Stephan H.; "Better Blast-Resistance Coming Soon to Façade", *Engineering News Record,* July 10, 2000, p 26.
Hoke, John Rat; *Architectural Graphic Standards,* Ninth Edition, Ramsey/Sleeper, John Wiley & Sons, Inc, New York, 1994.
Jodidio, Philip; *Santiago Calatrava,* Benedikt Taschen Verlag GmbH, Köln, 1998.
Nashed, Fred; *Time-Saver Details for Exterior Wall Design,* McGraw-Hill, New York, 1995.

Index

Illustration Credits

American Architectural Manufactures Association, AAMA *72*

J.C. Barry, University of Queensland, Australia *14 bottom*

Boundary Layer Wind Tunnel Laboratory, London, Ontario *120*

Image redrawn after detail on p 269 in David Button and Brian Pye; *Glass in Building*, Pilkington Glas Limited, Butterworth Heinemann, Oxford, 1993. *147 top left*

Richard Bryant/Arcaid *134*

Image redrawn after detail on p 137 of Andrea Compagno, *Intelligent Glass Façades: Material Practice Design*; Birkhäuser, Basel/Berlin/Boston, 1999 *85 left*

Images redrawn after details on p 107 of *Intelligent Glass Façades 96 bottom, 133 top left*

Construction Research Laboratory (CRL) *11, 24 right, 34, 59 bottom, 60 bottom, 71, 73 right, 76 top, 77, 106, 126 right, 142 right*

Dewhurst Macfarlane and Partners *122, 123, 135, 136 top*

Image redrawn after detail on p 40 of Brian Edwards, *Green Buildings Pay*, E & FN Spon, London, 1998 *91 left*

Gartner, drawing provided courtesy of Josef Gartner GmbH & Co. KG *86*

GE Polymershapes *138 right*

Kurt J. Goodfriend *139*

Ellis Herwig/Stock Boston *117*

John Linden, photo courtesy of Murphy/Jahn, Inc Architects *21*

Johnson Architectural Images, AJ1507; © Artifice, Inc. *65*

Jong Soung Kimm, FAIA *112 top*

Lohan Associates, Inc *29 top right, 30 top right*

Mary Margaret Loughran *59 top*

Murphy/Jahn, Inc Architects *20 top right*

NASA *82 top*

O'Malley Creadon Productions *61*

Prestressed/Prestressed Concrete Institute, PCI *38 bottom, 49 left*

Image redrawn after detail on p 27 of Reese, Nancy I.Z. and Phil Geib; "Danger From Above", *Chicago Tribune,* July 16, 2000 PPG Industries, Inc. *28*

Eric Rowland, AIA *115 top and center*

Image after drawing in Thomas A. Schwartz, "The John Hancock Tower Glass Failure – Debunking the Myths", *Proceedings of May 2000 ASCE Structures Congress,* Philadelphia, 2000 *119 bottom*

M. Shoemaker, Cesar Pelli & Associates *143*

Ezra Stoller; © ESTO *37*

Images redrawn after replacement system from Super Sky Products *68*

Wes Thompson provided by Enclos Corporation *148* right

Viracon *56 bottom, 111, 142 left* and *center*

Visionwall Corporation *96 center*

Image redrawn after detail on p 133 of Robert J. Kudder and Jeffrey L. Erdly, editors, *Water Leakage Through Building Façades*, ASTM, 1998 *70 below*

Jay Williams, AIA *118*

All other illustrations were provided by the author.

Design: Gabrielle Pfaff, Berlin

This book is also available in a German language edition.
(ISBN 3-7643-6713-X)

Bibliographic information published by Die Deutsche Bibliothek
Die Deutsche Bibliothek lists this publication in the Deutsche Nationalbibliografie; detailed bibliographic data is available in the Internet at <http://dnb.ddb.de>

P.O. Box 133, CH-4010 Basel, Switzerland
A member of the BertelsmannSpringer Publishing Group
Printed on acid-free paper produced from chlorine-free pulp. TCF ∞

Printed in Germany
ISBN 3-7643-6712-1

9 8 7 6 5 4 3 2 1 http://www.birkhauser.ch

Front cover: The broken light does not affect the structural performance of adjacent glass panels. Greenhouse Parc Citroën, Paris.